# Conference on Energy, Environment, Economy and Thermodynamics (COFRET 2021)

IOP Conference Series: Earth and Environmental Science
Volume 960

Bucharest, Romania
14 – 16 October 2021

**Editors:**

**Adrian Clenci**
**Monica Costea**
**Michel Feidt**
**Rodica Niculescu**

ISBN: 978-1-7138-4942-1
ISSN: 1755-1307

**Printed from e-media with permission by:**

Curran Associates, Inc.
57 Morehouse Lane
Red Hook, NY 12571

CURRAN ASSOCIATES INC. proceedings.com

**Some format issues inherent in the e-media version may also appear in this print version.**



Institute of Physics
Dirac House, Temple Back
Bristol BS1 6BE UK

Phone: 44 1 17 929 7481
Fax: 44 1 17 920 0979

techtracking@iop.org

**Additional copies of this publication are available from:**

Curran Associates, Inc.
57 Morehouse Lane
Red Hook, NY 12571 USA
Phone: 845-758-0400
Fax: 845-758-2633
Email: curran@proceedings.com
Web: www.proceedings.com

# TABLE OF CONTENTS

IOP Conf. Series: Earth and Environmental Science **960** (2022) 011001 doi:10.1088/1755-1315/960/1/011001

# Preface

First of all, COFRET is an acronym coming from the French expression *"**Co**lloque **Fr**ancophone en **E**nergie, **E**nvironnement, **E**conomie et **T**hermodynamique"*, meaning *Conference on Energy, Environment, Economy and Thermodynamics.*

Thus, COFRET is a scientific event initiated in 2002 by an enthusiastic Franch-Romanian group with the aim of organizing a conference on energy, the environment, the economy and thermodynamics every two years, once in France and another time in Romania. Professor Michel Feidt from the University of Lorraine-France, and Professors Stoian Petrescu and Vsevolod Radcenco from the "Politehnica" University of Bucharest-Romania are the founders of this international conference of French-speaking origin. Since its creation, COFRET has sought to bring together specialists from academia and industry to present their achievements and thus, to create opportunities for collaboration in the fields of energy, environment, economy and thermodynamics.

The 10th edition (www.cofret2021.eu) scheduled to take place on site in 2020 at the University of Pitești, Romania had to be postponed for a year due to the Covid-19 health crisis. Therefore, in the period 14 - 16 October 2021, the University of Pitești organized the 10th edition in Bucharest with the support of the University "Politehnica" of Bucharest. This edition also benefited from the support of French Society of Thermotechnics (SFT), Romanian Society of Thermotechnics (SRT), Society of Automotive Engineers of Romania (SIAR) and Society of Automotive Engineers of France (SIA). It is emphasized that the involvement of the SIAR and SIA began with the 7th edition organized in 2014 by Le CNAM in Paris and signifies a more targeted orientation of COFRET towards sustainable road mobility, which for several years has been the subject of important debates at the international level. In other words, since 2014, COFRET was more open to presentations on the themes related to sustainable road transport.

In COFRET, the accent falls on thermodynamics, which is a discipline that has provided us with two well-known laws: the first on the conservation of energy and the second on the irreversibility of the phenomena that govern our lives. The success of the discipline of thermodynamics is evident when one analyzes the industrial revolutions that have taken place over time. Then, the control of energy and its impact on the environment and on the economy are crucial aspects for human society, hence, our desire to further develop COFRET to adapt it to the stakes and challenges of the human society from today and from tomorrow.

Coming back to the 10th edition of COFRET, it was organized in hybrid mode: online and on site at the Central Library of University "Politehnica" in Bucharest. The presentations were given using the well-known online communications platforms, Teams, Zoom, Skype. Despite the pandemics, fifteen people participated on site: five from Romania, five from France, three from Bulgaria and two from Poland.

The papers included in this volume are the result of a double-blind review and we hope they will generate fruitful discussions about the exciting topics of Energy, Environment, Economy and Thermodynamics.

*The editors of COFRET2021 proceeding*

*Adrian CLENCI, Monica COSTEA, Michel FEIDT, Rodica NICULESCU*

| COFRET2021–SCIENTIFIC COMMITTEE | |
|---|---|
| Dumitru CHIRLEŞAN, University of Piteşti, Romania | CHAIRMEN |
| Adrian CLENCI, University of Piteşti, Romania | |
| Daniel ANGHEL, University of Piteşti, Romania | |
| Alexandru BOROIU, University of Piteşti, Romania | |
| Radu CHIRIAC, University « Politehnica » of Bucharest, Romania | |
| Alexandru CHISACOF, University « Politehnica » of Bucharest, Romania | |
| Carla CORADINI, USAMV Timisoara, Romania | |
| Monica COSTEA, University « Politehnica » of Bucharest, Romania | |
| Dinu COVACIU, University Transilvania of Braşov, Romania | |
| Mădălin CĂLBUREANU Popescu, University of Craiova, Romania | |
| Amélie DANLOS, Le Cnam Paris, France | |
| Michel FEIDT, University of Lorraine, France | |
| Daniela GIOSANU, University of Piteşti, Romania | |
| Ioana IONEL, Polytechnical University of Timişoara, Romania | |
| Monica IORDACHE, University of Piteşti, Romania | |
| Ivelina IVANOVA, University of Chemical Technology and Metallurgy, Sofia, Bulgaria | |
| Francois LANZETTA, University Franche-Comté, Belfort, France | |
| Emilia MLADIN, University « Politehnica » of Bucharest, Romania | |
| Emilian MADOSA, USAMV Timisoara, Romania | |
| Viorel NICOLAE University of Piteşti, Romania | |
| Rodica NICULESCU, University of Piteşti, Romania | |
| Eduard NIȚU, University of Piteşti, Romania | |
| Pierre PODEVIN, Le Cnam Paris, France | |
| Alexandru RACOVITZĂ, University « Politehnica » of Bucharest, Romania | |
| Alin RIZEA, University of Piteşti, Romania | |
| Monica SIROUX, INSA Strasbourg, France | |
| Ioana STĂNCULESCU, University of Bucharest, Romania | |
| Ştefan TABACU, University of Piteşti, Romania | |

### Peer review declaration

All papers published in this volume of **COFRET2021** have been peer reviewed through processes administered by the Editors. Reviews were conducted by expert referees to the professional and scientific standards expected of a proceedings journal published by IOP Publishing.

- **Type of peer review: ~~Single-blind /~~ Double-blind ~~/ Triple-blind / Open / Other (please describe)~~**
- **Conference submission management system: http://cofret2021.eu/index.php/cofret/2021/login**
- **Number of submissions received: 32 full papers (but more submissions received as abstracts)**
- **Number of submissions sent for review: 32**
- **Number of submissions accepted: 23**
- **Acceptance Rate (Number of Submissions Accepted / Number of Submissions Received X 100): 71.8%**
- **Average number of reviews per paper: 2**
- **Total number of reviewers involved: 28**
- **Any additional info on review process:**

# COFRET2021 Paper Review Form

This form is intended for full paper review.

*So, **first**, simply check the boxes that best express your opinion of the manuscript.*

### Manuscript type

- ○ Research Paper
- ○ Case study
- ○ Review/Overview Paper

**Abstract**
*Does it contain the essential information of the article? Is it complete? Is it suitable for inclusion by itself in an abstracting service?*

IOP Conf. Series: Earth and Environmental Science **960** (2022) 011002 doi:10.1088/1755-1315/960/1/011002

- Poor
- Marginal
- Acceptable
- Good
- Outstanding

**Originality**
*Does the paper contain new and significant information adequate to justify publication? Is the work relevant and novel?*

- Poor
- Marginal
- Acceptable
- Good
- Outstanding

**Relationship to literature**
*Does the paper demonstrate an adequate understanding of the relevant literature in the field and cite an appropriate range of literature sources? Is any significant work ignored?*

- Poor
- Marginal
- Acceptable
- Good
- Outstanding

**Organization of manuscript & Quality of Communication & Clarity of results/tables/illustrations**
*Does the paper clearly express its case/methodology, measured against the technical and scientific language of the field? Has attention been paid to the clarity of expression and readability, such as sentence structure, jargon use, acronyms, etc? Diagrams, figures, tables and captions: Are they essential and clear? Text and mathematics: Are they brief but still clear? If you recommend shortening, please suggest what should be omitted.*
*Are the results presented clearly and analysed appropriately? Do the conclusions adequately tie together the other elements of the paper? Are the conclusions summarizing what has been learned and why it is interesting and useful?*

- Poor
- Marginal
- Acceptable
- Good
- Outstanding

**English usage**
*We all should be aware that the standard of English/French is very important for an accurate assessment of the science presented in an international scientific article.*

IOP Conf. Series: Earth and Environmental Science **960** (2022) 011002 doi:10.1088/1755-1315/960/1/011002

- Poor
- Marginal
- Acceptable
- Good
- Outstanding

**Are you aware of prior publication of this work? or Have significant parts of the manuscript already been published?**

- Yes
- No

**Is the manuscript free of commercialism?**

- Yes
- No

***Secondly***, *please, suggest whatever modifications you think are necessary to improve the quality of the manuscript; if you feel it's the case, please, develop your choices from above.*

**Please, suggest (mandatory field):**

This is just an exemple:
The following should be explained since

***Finally***, *what is your recommendation for publication of the manuscript with IOP Publishing Ltd UK volume (http://conferenceseries.iop.org/) – indexed in Clarivate Web of Science and Scopus (mandatory field)*

- Accept
- Minor_revision
- Major_revision
- Reject

- **Contact person for queries: Adrian CLENCI, adi.clenci@upit.ro, adrian.clenci@lecnam.net**

# Innovation to zero emissions: nuclear energy trough nonproliferation and environment protection- Romanian experience for GEN 4

**M Valeca[1*], S Valeca[1] and D Giosanu[2]**

[1]University of Pitesti and Institute for Nuclear Research; 110142 Pitesti, Romania
[2]Department of Environmental Engineering, University of Pitesti, Targul din Vale Street, 110042, Pitesti, Romania

*e.mail: monica.valeca@nuclear.ro
ORCID: https://orcid.org/0000-0003-0648-3659

**ABSTRACT.** The UN Intergovernmental Panel on Climate Change (NY 2019) has concluded that nations must move more swiftly to reduce emissions of carbon dioxide, methane, and other greenhouse gases to avoid the most devastating effects of global warming. The paper presents nuclear energy as part of the solution. Due to the fact that the population is concerned about nuclear proliferation, plant safety and radiation protection, the paper presents the Romanian experience regarding the reduction of the risk of proliferation as well as the project of the 4th generation reactor ALFRED. One of the most important steps in assessing the candidate materials for Generation IV reactors is the material performance under neutron irradiation. In this respect, the paper also presents the results of the evaluations on some potential materials to be used in fast lead cooled reactors.

## 1. **Introduction**

The 2015 United Nations Climate Change Conference (COP 21) negotiated the Paris Agreement, a global agreement on the reduction of climate change, the text of which represented a consensus of the representatives of the 196 attending parties. On 22 April 2016 (Earth Day), 174 countries signed the agreement in New York [1], and began adopting it within their own legal systems.

The UN Intergovernmental Panel on Climate Change (NY 2019) has concluded that nations must move more swiftly to slash emissions of carbon dioxide, methane, and other greenhouse gasses to avoid the most devastating effects of global warming. The Climate Action Summit reinforced 1.5°C as the socially, economically, politically and scientifically safe limit to global warming by the end of this century, and net zero emissions by 2050 as the global long-term climate objective for all. Countries need to urgently accelerate work to define what this entails for the short-term (2020) and mid-term (2030) commitments that will be captured in their Nationally Determined Contributions and ensure the alignment of strategies to meet these commitments [2].

Climate changes are influenced by a number of factors, of which three are more important and can be expressed by the relationship below:

$$CO2 = P \times S \times E \times C$$

P = World population,
S = Services,
E = Energy efficiency
C = Carbon emissions

From the point of view of energy production depending on the fuels used, the amount of $CO_2$ released into the atmosphere is shown in table 1:

**Table 1.** The value of GHG emissions.

| The value of GHG emissions from different electricity generation technologies | GHG emissions, in tones $CO_2$ e/GWh |
|---|---|
| Lignite | 1054 |
| Coal | 888 |
| Oil | 733 |
| Natural gas | 499 |
| Biomass | 45 |
| ***Nuclear*** | ***29*** |
| Hydroelectric | 26 |
| Wind | 26 |

In this context, electricity generation from nuclear power plants is almost free of direct greenhouse gas emissions. IAEA found a range for emissions from the nuclear fuel cycle between 9 and 29 $CO_2$ e/GWh.

If the nuclear energy is part of the solution for reduction of GNG, is necessary to increase the acceptance for:

- Reducing the amount of weapon grade materials and support the nonproliferation
- Specific programs for radioprotection, includes: new technologies, new generation of nuclear reactors
- Reducing the nuclear waste volume
- Increase the nuclear safety
- Programs for plant life extension

## 2. Romanian achievements to support nonproliferation

*2.1 Full conversion of ICN Pitesti TRIGA 14 Mw Research Reactor core from HEU to LEU fuel and repatriation of HEU fuel in the country of origin*

- 1984 – Agreement with US-DOE – participation in "Reduced Enrichment in Research and Testing Reactors – RETR" programme and first activities performed in cooperation with US-ANL;
- 1990 – Establishment of the TRIGA-14 Mw core conversion research contract with US-ANL and supply of the TRIGA-LEU nuclear fuel sponsored by US-DOE third supply agreement USA-Romania
- 1991 – Delivery of the first 5 LEU fuel bundles (experimental fuel) by USA;

- 1991 – First LEU fuel load in TRIGA core, with US-ANL specialists' participation; the beginning of the physical conversion
- 996 – 2004 – TRIGA-14 Mw gradual core conversion by replacement of HEU spent fuel with fresh LEU fuel up to a core configuration with 35 fuel bundles.
- 1999 – Signing of the contract for the return of spent nuclear fuel used until 2009, according to the commitment to cease the use of HEU nuclear fuel on 12 May 2006. Contract between RENEL– GEN – ICN-Piteşti and US-INEL, and fully sponsored by USDOE;
- 1999 – Approximately 25% HEU spent fuel inventory was returned in the USA;
- 1999 – Approval of IAEA Project ROM/4/024: "Full Conversion of TRIGA-14 Mw from HEU to LEU"
- 2003 – Negotiation and signing of the tripartite contract IAEA – CERCA (France) – Romania for procurement of 16 LEU nuclear fuel manufactured in France. The financial support is granted by US-DOE and the Romanian Government;
- 2004 – 2006 delivery by CERCA- France of 16 LEU nuclear fuel in three shipments, to accomplish the full core conversion;
- 10 May 2006 – last time when HEU fuel was used;
- 12 May 2006 – deadline established by environmental agreement (EIS) for the return of the HEU type fuel used in research reactors (FRRSNFR);
- 15 May 2006 – HEU type nuclear fuel was transferred from research reactor to spent fuel storage system, to be prepared for return;
- 18 May 2006 – a new core configuration, fully converted, containing only LEU fuel was established;
- 24 May 2006 – first power rise for fully converted core operation-The 14 MW TRIGA research reactor

**Figure 1**. TRIGA HEU spent fuel management.

*2.2 Repatriation of Enriched Fuel from Romanian VVRS type Research Reactor-IFIN Măgurele in the Russian Federation (reactor shut down in 2002)*

- In 2002, after shutdown of the research reactor from IFIN-HH, the first important decommissioning action was the repatriation of HEU fresh fuel (83%) in the Russian Federation, which took place in 2003, being conducted by ICN - Pitesti;
- In 2004, the "U.S.-Romanian agreement on implementing nuclear non-proliferation projects" was signed between US-DOE and the Romanian Nuclear Agency, for the repatriation of high enriched fuel from the VVRS Reactor. Based on the press release: "The DOE- U.S.-Romanian agreement on implementing nuclear non-proliferation projects was signed Monday in Washington at the Romanian Consulate by Abraham and Romanian Minister Delegate of the Commission for Nuclear Energy S. Valeca". "This agreement provides yet another excellent opportunity for the United States and Romania to work together to reduce the threat of terrorism through the removal of proliferation-attractive material under the Global Threat Reduction Initiative", Abraham said Monday in a press statement.
- On 28 June 2009, based on the agreement between Romania and the Russian Federation the entire quantity of fuel burned S-36 containing HEU from the reactor at VVRS IFIN-HH was sent in the origin country by air in a single shipment; such operation represented a world premiere;
- On 1 November 2009, an amount of uranium metal enriched more than 20%, which was stored under control guarantees at ICN – Pitesti was sent to the Russian Federation;
- On 4 July and 23 December 2012, all LEU spent fuel at reactor VVRS was sent to the Russian Federation;
- Conducting these activities under non-proliferation programs meant a long series of diplomatic actions, negotiations in bilateral, trilateral or multilateral formats at the highest level, with a view to establishing the legal framework for import, export, transport and transfer of classified strategic materials under the supervision of IAEA safeguards and EURATOM;

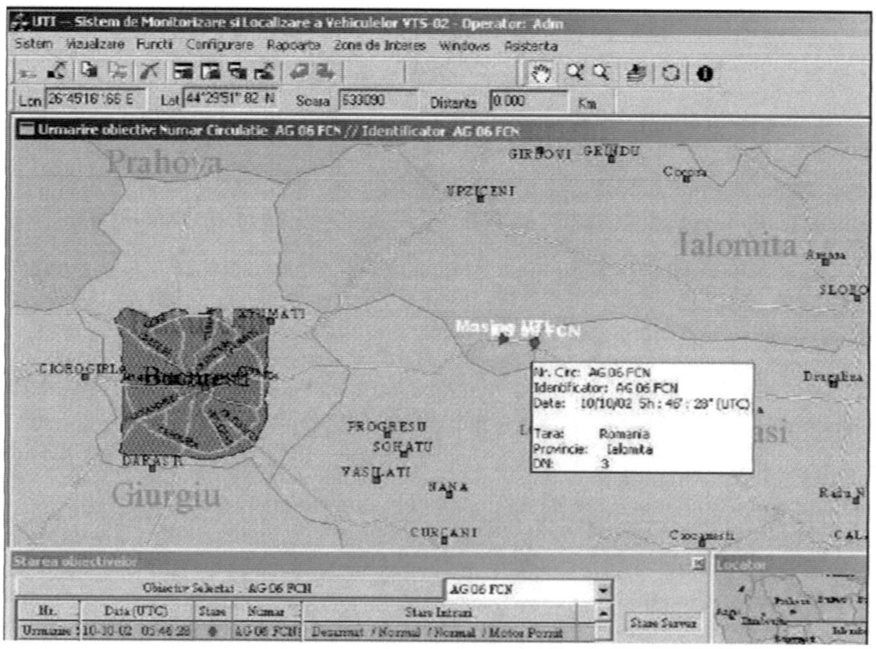


**Figure 2.** Satellite Monitoring of nuclear transport in Romania.

## 3. Decarbonization: Small modular reactors (SMR's) a possible response

Since 2001, the Generation IV International Forum (GIF) was created as a co-operative international endeavor seeking to develop the research necessary to test the feasibility and performance of fourth generation nuclear systems. Eight technology goals have been defined for Generation IV systems in four broad areas: sustainability, economics, safety and reliability, and proliferation resistance and physical protection. Small modular reactors (SMRs) can offer unique economic advantages that are not possible with large nuclear reactors. SMRs are designed in modules that can be fabricated in a factory and assembled on site [3].

One of the eight technologies selected is The Lead-cooled Fast Reactors (LFRs) feature a fast neutron spectrum, high temperature operation, and cooling by either molten lead or lead-bismuth eutectic (LBE), both of which support low-pressure operation, have very good thermodynamic properties, and are relatively inert with regard to interaction with air or water. Romania through RATEN-ICN formed two consortia: one international (FALCON) and one national (CESINA), for the development of the LFR demonstration reactor called ALFRED.

Decision to host in Romania the Advanced Lead Fast Reactor European Demonstrator (ALFRED) was based on:

*(a) Energy security*

-Energy Strategy 2016-2030: nuclear energy remains a strategic option for Romania.

-Further ambition: the whole future power production capacity based on advanced technologies, ensuring the stability and sustainability through high efficiency, flexibility and low carbon emissions.

*(b) Industrial development*

-nuclear industry strongly determined to maintain its competences and to continue to play a role in the future of the nuclear program.

-ALFRED as an opportunity to participate in the construction of the complex R&D infrastructure and of the demonstrator

*(c) Research, Development and Innovation*

-Synchronism: national RDI - European major themes and projects

-Boost the cooperation with the RDI major players

-Enlarge national research infrastructure

-Make nuclear attractive and keep young talents

*(d) Socio-economic development*

-create new full-time jobs and will maintain high expertise positions

-growing the economy

The applications of modular nuclear reactors are multiple. A studied application is the "smart city" which also includes the "smart company" in the category of large energy consumers. In this case, the discussions are made with the City of Mioveni and the "Dacia-Renault" Car Company, which is in the vicinity of ICN. The 100 MWe production of the ALFRED reactor ensures the needs of the three consumers mentioned above. Last but not least, a possible application is represented by the military field: providing electricity and thermal energy to isolated military units.

ALFRED is an essential step to reach the technology maturity level for the industrial implementation of the European Lead cooled Fast Reactor (ELFR).

## 4. Specific materials

Various stainless steels and other alloys have been considered and investigated during time as candidates for GEN IV and a large number of studies have been performed and reported regarding their behavior under different neutron spectra. Some desirable characteristics for the Gen-IV structural materials are:

(a) excellent dimensional stability against thermal and irradiation creep, void swelling, etc.;

(b) favorable mechanical properties such as strength, ductility, creep rupture, fatigue, creep–fatigue interactions, etc.

(c) acceptable resistance to radiation damage (irradiation hardening and embrittlement) under high neutron doses (10– 150 dpa or displacements per atom), etc;
(d) high degree of chemical compatibility between the structural materials and the coolant as well as with the fuel. In this regard, stress corrosion cracking (SCC), irradiation assisted stress corrosion cracking (IASCC) and many other issues are important.

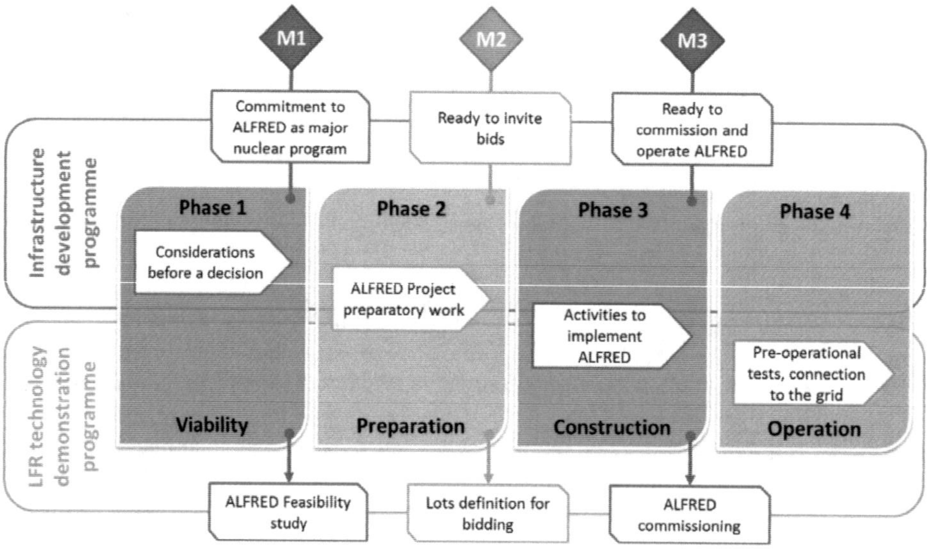


**Figure 3**. ALFRED chart.

Starting from the favorable characteristics and the potential for high performances of both Ti 15-15 and T91 stainless steels a preliminary analysis of their behavior under neutron irradiation in a Lead cooled Fast Reactor (the ALFRED Demonstrator) have been performed.

In this paper will be presented the results related to Ti 15-15. The reference active core configuration that has been designed and two computational tools have been used in the analysis of the two cladding materials: a) MCNPX 2.6.0 – for criticality simulations and evaluation of the neutron flux in the fuel claddings [4] and b) EASY-2010 (the European Activation System) that includes the burn up and inventory code FISPACT used to evaluate the activity, the isotopic inventory and the hazard factors for each cladding material based on the flux-spectrum computed by MCNPX.

*4.1Results for Ti 15-15 as fuel cladding material*

In order to evaluate the quantities of interest for Ti 15-15, the total flux in the fuel claddings (normalized to the reactor power) of the central fuel assembly has been evaluated using MCNPX 2.6.0 [5]. As next step the neutron flux spectrum has been used in the FISPACT code in order to compute the activation of Ti 15-15 after 5 years of irradiation. Its activity, at various cooling time steps, in order to analyze its classification as waste, as well as the doses and the hazard factors for ingestion and inhalation is presented in table 2.

**Table 2.** Summary of Ti 15-15 activation (1 kg of Ti 15-15).

| | Time | Activity (Bq) |
|---|---|---|
| Irradiation steps | 1 year | 3.3820E+13 |
| | 5 years | 3.8180E+13 |

| | | |
|---|---|---|
| Cooling steps | 1 day | 1.9973E+13 |
| | 1 year | 4.8158E+12 |
| | 10 years | 5.4284E+11 |
| | 100 years | 3.3542E+10 |
| | 1000 years | 3.2401E+09 |
| | 10 000 years | 1.3070E+09 |
| | 100 000 years | 3.6360E+08 |

FISPACT code results show that the total activity of T91 (1 kg) after 5 years of irradiation is almost of the same magnitude (~ 1013 Bq, slightly lower) as of Ti 15-15. Similar to Ti 15-15, the activity, doses and the inhalation and ingestion factors have still high values after 100 years of cooling.

It should be mentioned that the errors (for both cross sections and half-lives) associated to the FISPACT code computations of the total activity and dose rate values could be around 10% for short irradiation and cooling times and will become much higher for very long cooling times (hundreds of years). Another aspect that could be of interest especially for the nuclear materials field is the behavior of the Ti 15-15 isotopic composition with irradiation.

## 5. Conclusions

In order to meet the requirements to decrease the greenhouse emissions, nuclear energy is part of the solution for Romania.

The existing Romanian experience and the continuous decision of nonproliferation and the task to decrease in the volume of radioactive waste and potential accident to protect the population and environment is projected trough Generation 4 reactors, SMR type.

The efforts of the Romanian scientific community for the development of the ALFRED reactor are part of the international effort. The analysis of materials and nuclear safety made by ICN specialists are part of this effort.

The preliminary analysis performed provides information about the behavior of the fuel cladding materials under irradiation as well as on the aspects (activities, doses, hazard factors) regarding the personnel radioprotection. It should be also emphasized that the transmutation processes that may occur in some structural materials makes more difficult the accurate evaluation of the neutron irradiation damage which value is one of the main constraints during the design stage of a reactor.

Exploring beyond the boundaries of the field based on education and research will lead to the development of new high-performance technologies in Romania.

## 6. References

[1] Falk P 2016 *U.S. joins 174 nations to sign hard-won climate pact.* CBSNEWS.
[2] Findlay T 2010 *The future of nuclear energy to 2030 and its implications for safety, security and nonproliferation,* Part 1 – The Future of Nuclear Energy to 2030, The Centre for International Governance Innovation (CIGI), Waterloo, Ontario, Canada
[3] IAEA, 2011, Workforce planning for new nuclear power programmes
[4] Petrovich C, Grasso G, Artioli C, Rocchi F and Sciora P 2012, Definition of the ETDR core and neutronic characterization, DEL007-2012, WP2, LEADER project, UTFISSM-P9SZ-002, Rev.0, ENEA-CEA.
[5] Pelowitz D B et al. 2011 MCNPX 2.7.E Extension, LA-UR-11-01502, Los Alamos.

IOP Conf. Series: Earth and Environmental Science **960** (2022) 012002 doi:10.1088/1755-1315/960/1/012002

# The identification of optimal operating parameters under high energy efficiency conditions for a single polymer electrolyte membrane (PEM) fuel cell

I G Bratu[1], R F Ene[2], M Vulpe[1], F Uleanu[1] and D Giosanu[1]
[1)] University of Pitești, Targu din Vale Street, No. 1, 110040 Pitești, Romania
[2)]National Research and Development Institute for Cryogenic and Isotopic Technologies, Uzinei Street No. 4, 240050, Rm. Valcea, Romania

*e-mail : gabriel.bratu@upit.ro
ORCID : https://orcid.org/0000-0002-2398-2946

***Abstract:*** *The performance of PEM fuel cells is influenced by several factors such as: the operating temperature of the cell, the reactant gas flow, work pressures, the reaction gas humidity. In the present work we aimed to identify the optimal values of these parameters for operation of a PEM cell to achieve maximum power in conditions of high efficiency; the technological possibilities of its use in a portable energy application have been evaluated. Experimental measurements regarding the integrating polymeric membrane in three different fuel cell construction designed were performed. The influence of the mechanical compression of the GDL diffusion layer on the total internal resistance of the cell was achieved by comparative analysis of the polarization curves. It was found that as the deformation level of the MEA increases, the power generated by the battery increases progressively. The resulting experimental data subsequently allowed the design and implementation of a PEM fuel cell assembly, fully functional at power level, corresponding to the number of constituent elements.*

## 1. Introduction

In order to be exploited, especially in the development stage, a PEM fuel cell needs a control of the input parameters (pressures, flow rates, temperatures, humidity), as well as the output parameters (temperature, voltage, current) [1]. This is done with the help of an installation that has the ability to manage pneumatically, humidified and pressurized gas flows at variable temperatures, but also controls the electrical parameters discharged, the current extracted at a certain voltage [2].
In order to manage a power of the order of watts, it is necessary that in the installation there are flow regulators / controllers of the order of milliliters-liters / minute. Also, the management of such flows is done on appropriate technological routes, i.e. on hoses/pipes with larger and larger sections, corresponding to the flows proportional to the power range [3].
The complexity of the fuel cell test system increases as the working range increases and the control of several parameters is desired, as well as the way the battery is operated (load, thermal regime, response times of the control device) [4].

## 2. Experimental determinations

The experimental determinations on the single PEM battery (figure 1), that will be considered the battery with the reference values in the design of the battery assembly for the work application were performed on an own construction installation, which can manage an electric power in the range 1W-500W, with working parameters adjustable (stoichiometry, pressures, humidity, temperature) [5].

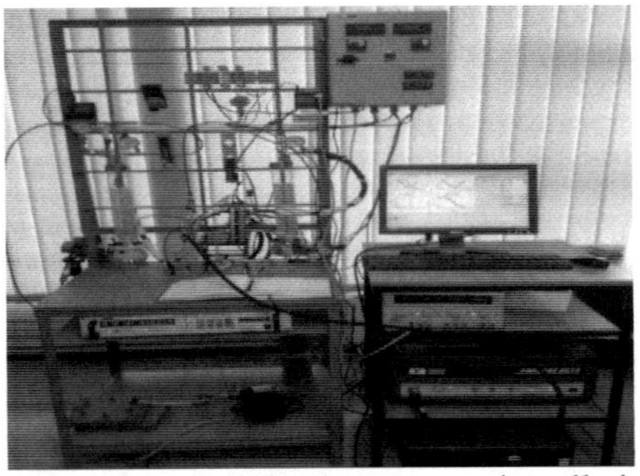

**Figure 1**. Single PEM battery test installation.

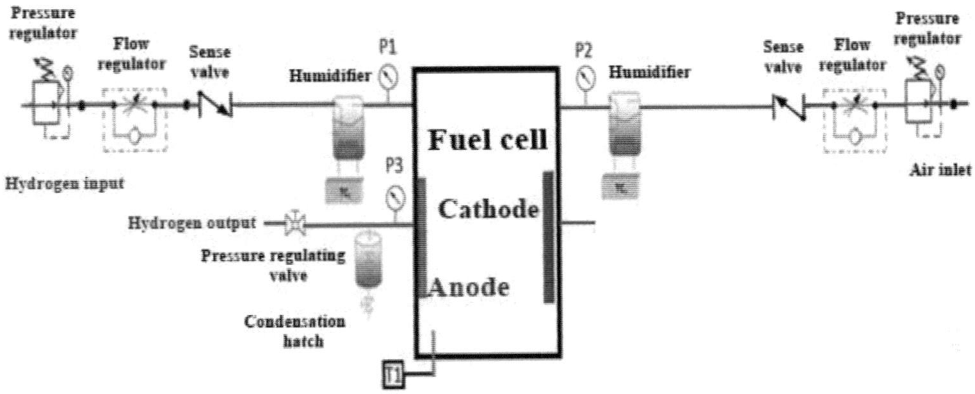


**Figure 2**. Schematic-block experimental installation for single file.

The installation includes (figure 2):

- Michel PC Mini 70 humidity sensor that monitors the relative humidity level of the cathode. It is mounted inline on the cathode path after the humidification hatch, in a special T-type mounting, just before the gas enters the fuel cell; In parallel, a Lutron HT 3009 Thermohygrometer was also used to increase the accuracy of the measurements (figure 3).

**Figure 3**. Humidity sensors of gas flows.

- water condensing hatch on the anodic route, mounted at the outlet of the pneumatic installation, to collect the discharged water by separating the liquid / gaseous phase, for the purpose of various volume measurements of the water produced, to measure its acidity, but also to remove water from exhaust flow, gas that will be measured later, with flow meters whose reading /operation is negatively influenced by the humidity on the routes;
- electronic load AMREL PLA 800-30-300 for simulating operating modes and tracing the polarization curve;

- 3 manometers indicating the gas pressures before the humidification hatch and at the level of the cell, for each compartment;
- directional valves that prevent the return of gas from the humidification hatches when the supply is stopped; are mounted at the output of the flow regulators;
- a control valve used to create the desired values for the working pressures in the anodic compartment. For the cathode compartment the air outlet is free, the working pressure being generated only by the value of the supply flow and by the pneumatic resistance of the pipes at the level of the plates;
- connecting elements specific to the pneumatic field, Swagelok type;
- technological routes from pressure and temperature resistant PVC hose, thermally insulated and heated stainless steel metal routes to prevent water condensation on the humidifier-fuel cell distance;
- a pressure regulator, which adjusts to the desired value on the hydrogen routes;
- 2 mass and volumetric flow regulators, mounted on each gas path, at the inlet to the cell, in order to supply the fuel cell to the desired flows / stoichiometries;
- 2 humidifiers, one on each route, mounted at the entrance to the battery, in order to humidify the inlet gases; In this case, only the one on the cathodic route was used, as the humidity of the hydrogen that will be produced by the generator in the laboratory is not yet known. They are self-built, bubbling type, made of plexiglass, equipped with gas diffuser, heating element and temperature regulator, with a height of 500 mm and an inner diameter of 56 mm. The change of the relative humidity is made, up to a certain desired degree and at a certain gas flow, either by changing the height of the water column, or by changing the temperature of the water in the hatch.

For the development of the single file considered the file with reference values, which will be the basis for the design and construction of a set of files, the entire testing procedure was customized. As the application itself ultimately provides for the portability of power generating equipment, this situation involves "access" to areas with different humidities, so that the condition of relative humidity of the reaction air was taken into account and experienced during the development of the single cell [10]. For example, the average annual air humidity in the Sahara Desert is 25%, in the Amazon jungle it is 95%, and in our country it is 55%.

Since the supply of a single cell with lightly pressurized air in the conditions of ambient humidity in the laboratory cannot be done easily, various degrees of relative humidity were simulated by humidifying the air flow supplied to the cathode by means of the bubble humidifier. Mainly, for the tests performed in this work, tests were performed with relative humidity of the air fed to the cathode, between 55% and 95%.

**Figure 4**. Humidifier made for humidifying the air flow.

There are limitations regarding the temperature of the water in the humidification hatches, these being given by the maximum allowable temperature (70 ° C) for this type of plexiglass, under the pressure conditions at which the determinations were made (0-0.3bar).

The development of a PEM-type battery pack that can be fitted into a portable and energy-efficient equipment also involves, in addition to a battery pack with high power densities relative to mass and

volume, the use of a fan that provides the air required for the cathodic reaction, energy efficient [6]. Its choice has an essential role in the final efficiency of the equipment produced, being conditioned, in addition to the flow rates and pressures and energy consumption. To supply the single cell, an air distribution grid has been designed and built that allows the airflow to be directed as evenly as possible, without large pressure drops, laterally, towards the cathodic channels.
The amount of air supplied to the cathode of the single cell was ensured by a compressor, its flow and pressure being controlled by means of the pressure regulator, but also of the flow regulator integrated in the test stand [7].

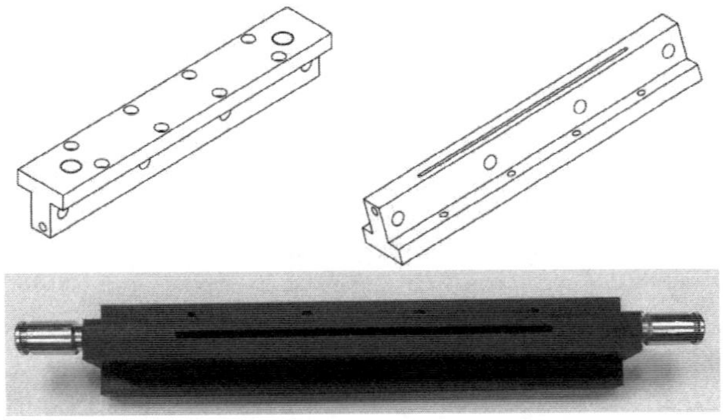

**Figure 5**. Air distribution grid at the cathode ducts.

The thermal management system of the fuel cells is also a key element of the test installations, a fuel cell not being able to function optimally, safely and for a long time, without a proper thermal control. The higher the range of power generated, the more heat must be managed.
For cooling PEM type fuel cells, various cooling techniques are used, some more practical, others quite rarely integrated.

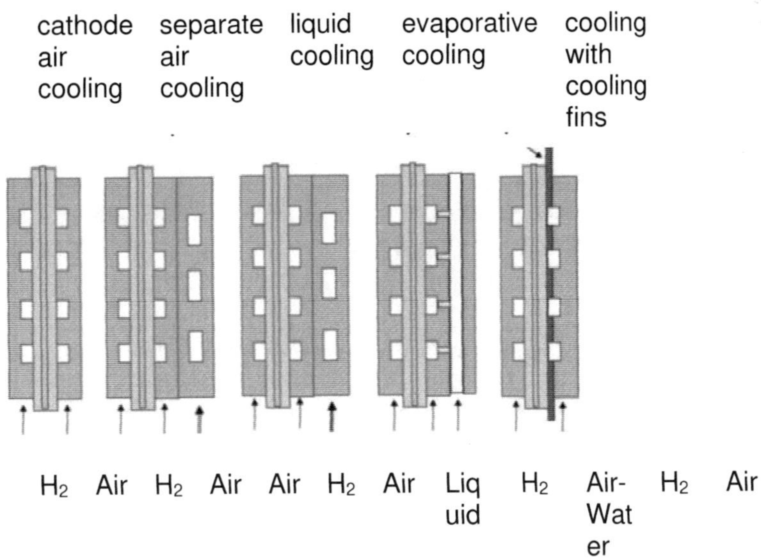


**Figure 6** - Types of cooling in PEM type fuel cells

-Cooling with supply air, when cooling can only be achieved with the air flow from the cathode. This type of cooling is only suitable for small battery assemblies, when usually its outer surface is larger than its active surface (natural air supply batteries, low pressure, so that the battery assemblies are capable to dissipate heat at a rate comparable to the rate of heat generation. A disadvantage is that it requires relatively large dimensions of the cathode channels compared to the anodic ones, automatically leading to an increase in volume of the built assembly.

- Separate compartment air cooling is used for battery assemblies that discharge powers of the order of hundreds of watts, so that the air is blown to remove the heat generated by electrochemical processes. This type of cooling can be used when volumetric power density is not a major factor, especially in stationary applications.
- Cooling with coolants can be done using deionized water, glycols or liquid solutions with suitable electrical properties. The thermal properties of liquids, such as specific heat capacity, thermal conductivity, are several orders of magnitude larger than in the case of gases, so that for high thermal loads, the choice of a liquid cooling system instead of air is the solution. natural.

Cooling can be arranged at the level of each cell, between pairs or groups of cells (this is only feasible for low power densities, as it will lead to high temperatures inside the battery pack).

An equal distribution of the coolant at the active areas of the membranes can be done through internal distribution routes, in the same way as the reactant gases.

- Evaporation cooling (phase change cooling) is used when the enthalpy of water vaporization is involved in cooling, by injecting additional water into the reactant flows of the PEM fuel cell. The heat generated at the electrodes is removed from the water by its phase change, keeping the temperature constant. This cooling method is also used to prevent drying of the polymer membrane
- Cooling with heat distributors, also called passive cooling, is based on the heat conduction in the direction of the plate plane and removes it from the central area to the edge of the bipolar plate. A special challenge in this type of cooling is represented by the values of thermal conductivity in the plane of the plates, values that must be very high, in order to control the temperature variations at the level of the active surfaces of the membranes.

## 3. Conclusions

Of the cooling techniques listed above, cooling with coolants is the most widely used technique by manufacturers of battery assemblies, being the most efficient, for powers greater than 2 kW or for areas of active surfaces larger than 50 $cm^2$.

Given the size of the active surface of the assembly, the electrode membrane used in the experiments, the current densities generated at the surface of the membrane, but especially the fact that the constructive solution of the battery assembly must provide high energy densities in volume and mass, the use of the battery cooling scheme with the help of the air supplied to the cathode was chosen.

## 4. References

[1] Barbir F 2005 *PEM Fuel Cells: Theory and Practice* (Elsevier Academic Press)

[2] Ebrașu D, Petreanu I, Pătularu L, Ștefănescu I and Văleanu M 2008 *Design and Characterization of styrene-based proton exchange membranes* in *Functionalized Nanoscale Materials, Devices and Systems*, Part of the NATO Science for Peace and Security Series B: Physics and Biophysics book series (NAPSB)

[3] Hacquard A 2005 Thesis *Improving and Understanding Direct Methanol Fuel Cell (DMFC) Performance*, (Worcester Polytechnic Institute)

[4] Litster S and McLean G 2004 *Journal of Power Sources* **130** 221

[5] Ștefănescu I (coord) 2010 *Combustion cells - between theory and practice* (Conphys Publishing House, Rm. Vâlcea)

[6] Karimi S, Fraser N, Roberts B and Foulkes R F 2012 *Advances in Materials Science and Engineering*, Article ID 828070

[7] Zerouala M, Ben Moussab H and Tamerabet M 2012 *Energy Procedia* **18** 317-26

# Influence of boundary conditions on the optimization of a geothermal heat pump studied using a thermodynamic model

**A Arz[1], A Minghini[1], M Feidt[1], M Costea[2] and C Moyne[1]**

[1]L.E.M.T.A., Lorraine University, 2 avenue de la Forêt de Haye, 54516 Vandoeuvre-Lès-Nancy, France
[2]Engineering Thermodynamics Department, University Politehnica of Bucharest, 313 Splaiul Independentei, 060042 Bucharest, Romania

E-mail: monica.costea@upb.ro, https.orcid.org/0000-0002-0954-7286

**Abstract.** This paper is the logical follow-up to a work [1] whose results were presented at the 28th French Thermal Congress which was to be held in Belfort in 2020. The model developed at that time is completed in this proposal to consider the specificity of the geothermal heat pump. This is a machine operating upon a mechanical vapor compression cycle, the limit of which is an inverse Carnot cycle. Its specificity consists of a cold loop at the source with the geothermal exchanger and the evaporator, then a hot loop at the sink with the condenser and a floor heat exchanger in the application considered here. We are particularly concerned with the optimal sizing of these heat exchangers through their effectiveness. The parametric sensitivity of this distribution to various boundary conditions is studied, especially by focusing on different conditions at the source: (1) imposed soil temperature, corresponding to a Dirichlet condition, (2) imposed heat flux (including adiabatic case), corresponding to a Neumann condition, (3) imposed mechanical power consumed by the heat pump, and (4) imposed coefficient of performance COP, to all cases being associated a finite thermal capacity in thermal contact with the geothermal exchanger operating in steady-state conditions.

## 1. Introduction

### *1.1. First models of reverse cycle machines*

The first models of reverse cycle machines were developed based on equilibrium thermodynamics and they describe machines with two thermal reservoirs. These reservoirs are in fact thermostats, namely source and sink of infinite thermal capacity and therefore, temperatures $T_{CS}$ = const will be associated to the source and respectively $T_{HS}$, to the sink [2].

The simultaneous use of energy and entropy balances for a reverse Carnot cycle (refrigerating machine-RM, or heat pump-HP) leads to a limiting value of the First Law efficiency for the machine in the reversible case (according to figure 1: cycle 1-$2_{is}$-$3_{is}$-4). The energy balance on a machine cycle is written as:

$$W = Q_{HS} - Q_{CS} \tag{1}$$

where $Q_{HS}$, thermal energy output from the hot loop,
$Q_{CS}$, thermal energy input in the cold loop,
$W$, mechanical energy expense.

The entropy balances on the cycle expressed by using the entropy method is:

$$\frac{Q_{HS}}{T_{HS}} = \frac{Q_{CS}}{T_{CS}} + \Delta S_I \tag{2}$$

with $\Delta S_I$, entropy production due to internal irreversibilities of the cycle.

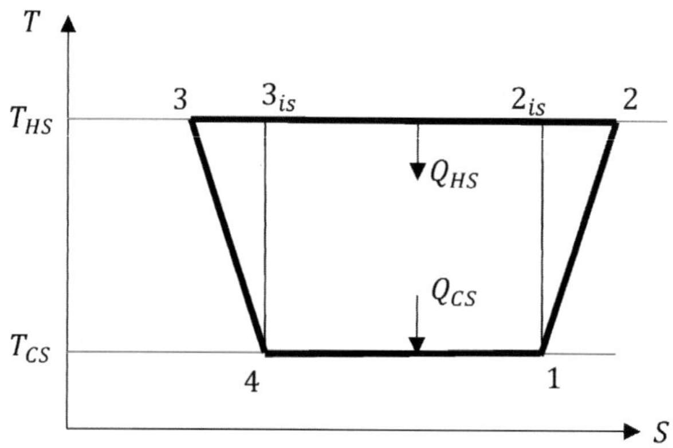


**Figure 1.** Reverse Carnot cycle: endo-reversible (1-$2_{is}$-$3_{is}$-4) or endo-irreversible (1-2-3-4).

As the study focuses on heat pumps, we recall here the coefficient of performance expression for a heat pump

- at imposed mechanical power $\dot{W}_0$:

$$COP_{PAC} = COP_{PAC_{rev}}\left(1 - \frac{T_{CS}\,\dot{S}_I}{\dot{W}_0}\right) \tag{3}$$

- at imposed calorific power $\dot{Q}_{H_0}$:

$$COP_{PAC} = COP_{PAC_{rev}}\left(1 - \frac{T_{CS}\,\dot{S}_I}{\dot{Q}_{H_0}}\right) \tag{4}$$

*1.2. The model of the geothermal heat pump*

The model of the geothermal heat pump is an advanced version of the previous scheme, comprising a cold loop and a hot loop, as illustrated in figure 2. It is assumed that these two loops operate adiabatically (without thermal losses) so that in steady-state regime one gets:

$$\dot{Q}_g = \dot{Q}_{evap}\ ;\ \dot{Q}_{cond} = \dot{Q}_p \tag{5}$$

with $\dot{Q}_g$, thermal power of the geothermal heat exchanger,

$\dot{Q}_{evap}$, thermal power of the evaporator,

$\dot{Q}_{cond}$, thermal power of the condenser and

IOP Conf. Series: Earth and Environmental Science **960** (2022) 012003 doi:10.1088/1755-1315/960/1/012003

$\dot{Q}_p$, thermal power of the heating floor at the hot end.

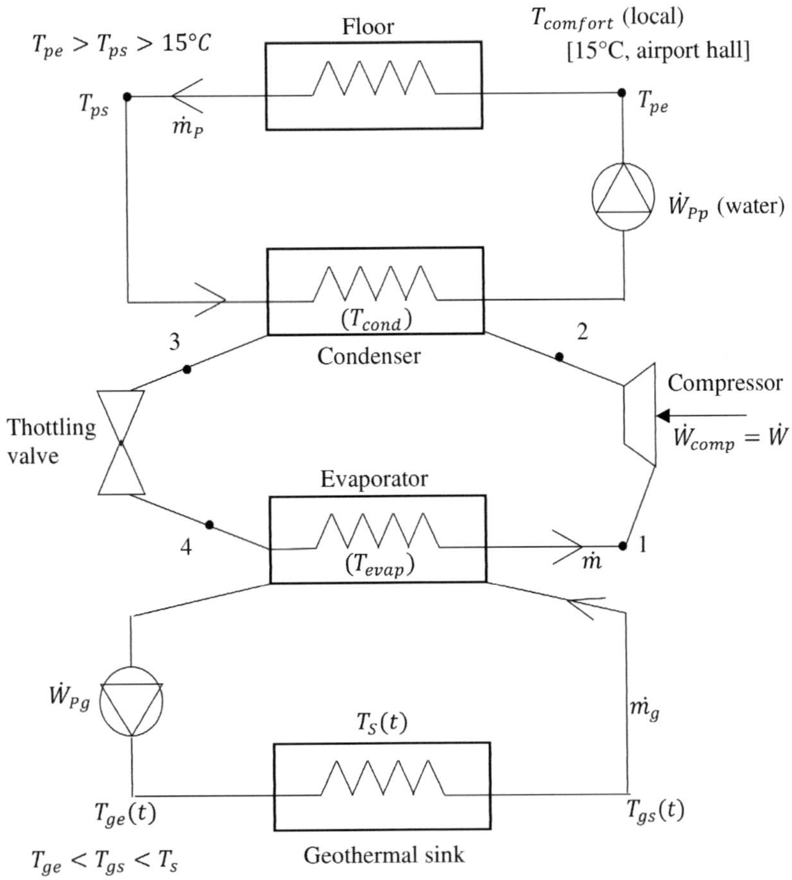


**Figure 2.** Theoretical scheme of a heat pump model in simple configuration.

The temperatures relative to this configuration and their variation ranges are shown in figure 2, together with the mass flow rates and thermal capacity rates of cold loop fluid $\dot{m}_g$, $\dot{c}_g = \dot{m}_g c_{p_g}$, then of hot loop fluid $\dot{m}_p$, $\dot{c}_p = \dot{m}_p c_{p_p}$. We notice a formal symmetry between the hot end and the cold end loops.

*1.3. The influence of finite physical dimensions on the behavior of the heat pump*

If we assume the geothermal source of infinite extension [3] the cylindrical geometry of the geothermal sink admits an isothermal type limiting condition:

$$\lim_{R\to\infty} T_s(R) = T_s$$

where $T_S$ is the soil temperature.

If we observe the practical realization (example of Zurich airport), the presence of a set of sources under the building leads rather on a square mesh to a limiting condition of adiabatic type $\left(-\lambda_s \frac{\partial T_s}{\partial x} = 0\right)$. We assume that this condition is well approximated by keeping the cylindrical geometry of the source. The heat pump itself can introduce finiteness conditions if we assume for example the imposed mechanical power $\dot{W}_0$. We can also consider an imposed heat extraction flow $\dot{Q}_0 = \dot{Q}_{evap}$, which leads to impose a constant heat flow from the geothermal source.

One of the objectives of this study is to quantify the influence of boundary conditions on the behavior of the heat pump (sizing and transient behavior over a heating season). Whatever the geothermal exchanger used (single tube, U tubes, bayonet exchanger), four cases will be considered: $T_S$ = ct; $\dot{Q}_0$ imposed; $\dot{W}_0$ imposed, and $COP_0$ imposed, with the consequences on a possible optimal dimensioning of the heat pump, through a study of parametric sensitivity.

## 2. Simplification of the study to the Chambadal model

### 2.1. The geothermal heat pump of Chambadal

Figure 3 illustrates the simplified scheme of the Chambadal geothermal heat pump.

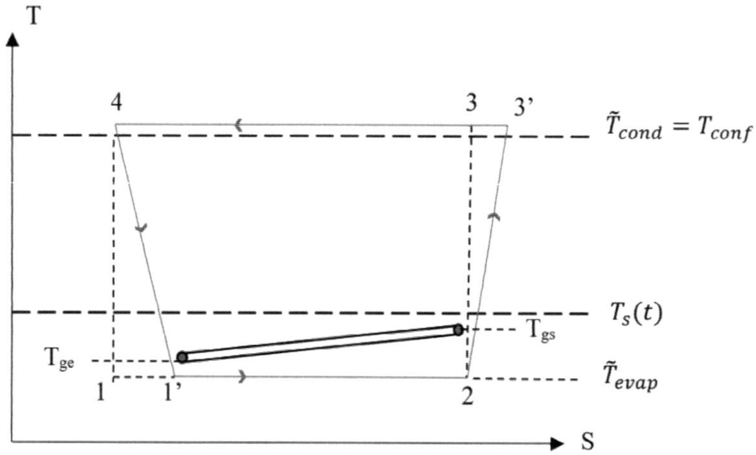


**Figure 3.** Diagram of the Chambadal geothermal heat pump.

It can be seen in this figure that the high temperature loop is assumed to be perfect, and therefore the set point temperature $T_{conf}$ is identical to the temperature of the condenser.

#### 2.1.1. Cases where $T_S$ is assumed constant [2]

At imposed flow rate and geometry, one gets:

$$\dot{Q}_g = \varepsilon_g \dot{c}_g (T_s - T_{ge}) \quad (6)$$

where $T_{ge}$, temperature of the geothermal source fluid at the inlet of the heat exchanger,
$\varepsilon_g$, effectiveness of the geothermal source exchanger.

The temperature $T_{ge}$ appears from (6) as a control variable. Then, one can show that the average evaporator temperature is:

$$\tilde{T}_{evap} = T_{ge} - (T_s - T_{ge})\varepsilon_g \left(\frac{1}{\varepsilon_{evap}} - 1\right) \quad (7)$$

The energy balance and the entropic balance of the endo-reversible cycle provide:

$$\dot{Q}_{cond} = \dot{Q}_g + \dot{W} \quad (8)$$

$$\frac{\dot{Q}_g}{\tilde{T}_{evap}} = \frac{\dot{Q}_{cond}}{T_{conf}} \quad (9)$$

Then, the mechanical power expense results in:

$$\dot{W} = \varepsilon_g \dot{c}_g (T_s - T_{ge}) \left[\frac{T_{conf}}{T_{ge} - (T_s - T_{ge})\varepsilon_g\left(\frac{1}{\varepsilon_{evap}} - 1\right)} - 1\right] \quad (10)$$

with $\varepsilon_{evap}$, effectiveness of the evaporator.

The expression of $\dot{W}$ appears as an increasing function of ($T_S - T_{ge}$), decreasing of $T_{ge}$, increasing of $\varepsilon_g$, $\varepsilon_{evap}$, and with the finite physical dimension constraint of $\varepsilon_g$, + $\varepsilon_{evap} = \varepsilon_g \leq 2$, an optimal distribution should result.

*2.1.2. Cases where $\dot{Q}_g$ is assumed constant $(\dot{Q}_0)$*

Unlike the isothermal condition (or Dirichlet condition) imposed in the previous subsection, this second condition corresponds to an extensive or Neumann condition. Hence it comes:

$$T_{ge} = T_s - \frac{\dot{Q}_0}{\varepsilon_g \dot{c}_g} \tag{11}$$

$$\tilde{T}_{evap} = T_s - \frac{\dot{Q}_0}{\dot{c}_g}\left(\frac{1}{\varepsilon_g} + \frac{1}{\varepsilon_{evap}} - 1\right) \tag{12}$$

The combination of the energy and entropy balances with the two preceding expressions leads to:

$$\dot{W} = \dot{Q}_0 \left[\frac{T_{conf}}{T_s - \frac{\dot{Q}_0}{\dot{c}_g}\left(\frac{1}{\varepsilon_g} + \frac{1}{\varepsilon_{evap}} - 1\right)} - 1\right] \tag{13}$$

This function is increasing by $\dot{Q}_0$ and decreasing by $T_S$. But for $T_S$ and $\dot{Q}_0$ parameters, it appears that the effectiveness finiteness constraint leads, under endo-reversible conditions, to the distribution of $\varepsilon_g$, $\varepsilon_{evap}$ related by:

$$\dot{\varepsilon}_g^* = \dot{\varepsilon}_{evap}^* = \frac{\varepsilon_T}{2} \tag{14}$$

The minimum mechanical power expense under endo-reversible conditions becomes:

$$min\ (\dot{W}) = \dot{Q}_0 \left[\frac{T_{conf}}{T_S - \frac{\dot{Q}_0}{\dot{c}_g}\left(\frac{4}{\varepsilon_T} - 1\right)} - 1\right] \tag{15}$$

*2.1.3. Cases where $\dot{W}$ is assumed constant $(\dot{W}_0)$*

The combination of the energy and entropy balances with the imposed mechanical power $\dot{W}_0$ constraint provides:

$$\dot{Q}_g = \dot{W}_0 \frac{\tilde{T}_{evap}}{T_{conf} - \tilde{T}_{evap}} \tag{16}$$

where the average temperature over the evaporator has the expression given in section 2.1.2:

$$\tilde{T}_{evap} = T_{ge} - \frac{\dot{Q}_g}{\dot{c}_g}\left(\frac{1}{\varepsilon_{evap}} - 1\right) \tag{17}$$

Equation (17) depending on $\dot{Q}_g$ has to be solved if $T_{ge}$ is a control variable. Otherwise, if $T_S$ is imposed by the source, the corresponding equation to solve will be:

$$\tilde{T}_{evap} = T_s - \frac{\dot{Q}_g}{\dot{c}_g}\left(\frac{1}{\varepsilon_g} + \frac{1}{\varepsilon_{evap}} - 1\right) \tag{18}$$

Then, equation (16) becomes a new second order equation expressed by:

$$\dot{Q}_g = \frac{(T_{conf} - T_S) + \frac{\dot{W}_0}{\dot{c}_g}\left(\frac{1}{\varepsilon_g} + \frac{1}{\varepsilon_{evap}} - 1\right) + \sqrt{\Delta}}{\frac{2}{\dot{c}_g}\left(\frac{1}{\varepsilon_g} + \frac{1}{\varepsilon_{evap}} - 1\right)} \tag{19}$$

with

$$\Delta = \left[\left(T_{conf} - T_S\right) + \frac{\dot{W}_0}{\dot{c}_g}\left(\frac{1}{\epsilon_g} + \frac{1}{\epsilon_{evap}} - 1\right)\right]^2 + \frac{4\dot{W}_0}{\dot{c}_g} T_S \left(\frac{1}{\epsilon_g} + \frac{1}{\epsilon_{evap}} - 1\right)$$

A possible approximation is provided by the following expression:

$$\dot{Q}_g \approx \dot{W}_0 \frac{T_S}{\left(T_{conf} - T_S\right) + \frac{\dot{W}_0}{\dot{c}_g}\left(\frac{1}{\epsilon_g} + \frac{1}{\epsilon_{evap}} - 1\right)} \quad (20)$$

Note that under the endo-reversibility and approximation conditions, the maximum of $\dot{Q}_g$ is obtained at the effectiveness equipartition:

$$Max\left(\dot{Q}_g\right) \approx \dot{W}_0 \frac{T_S}{\left(T_{conf} - T_S\right) + \frac{\dot{W}_0}{\dot{c}_g}\left(\frac{4}{\epsilon_T} - 1\right)} \quad (21)$$

*2.1.4. Case where COP is imposed ($COP_0$)*

One easily finds that $COP_0$ of the heat pump that is given by:

$$COP_0 = \frac{T_{conf}}{T_{conf} - T_{evap}} \quad (22)$$

with

$$T_{evap} = T_S - \frac{\dot{Q}_g}{\dot{c}_g}\left(\frac{1}{\epsilon_g} + \frac{1}{\epsilon_{evap}} - 1\right) \quad (23)$$

By combining equations (22) and (23) and setting $z_0 = \frac{COP_0 - 1}{COP_0}$, the resulting expression of the objectif function $\dot{Q}_g$ is:

$$\dot{Q}_g = \dot{c}_g \frac{T_S - z_0 T_{conf}}{\frac{1}{\epsilon_g} + \frac{1}{\epsilon_{evap}} - 1} \quad (24)$$

Like the result of the previous case, under endo-reversibility conditions, the maximum useful effect corresponds to the maximum of $\dot{Q}_g$ at the equipartition of the effectiveness:

$$Max\left(\dot{Q}_g\right) = \frac{\dot{c}_g}{\frac{4}{\epsilon_T} - 1}\left(T_S - z_0 T_{conf}\right) \quad (25)$$

Equation (25) also implies an inequality condition on $T_S$ expressed by $T_S \geq z_0\, T_{conf}$.

## 3. Discussions of the Chambadal geothermal heat pump model results

A parametric sensitivity study and the comparison of the 4 previous operating case results when the effectiveness finiteness condition of the geothermal source is considered, will be presented in this section.

### *3.1. Characteristic of the geothermal heat pump*

In addition to the existence of two loops (hot and cold), the geothermal heat pump has a transient source on a seasonal scale (heating or cooling) and moreover, it is finite so that:

$$\dot{Q}_g(t) = -M_s Cp_s \frac{d\bar{T}_s(t)}{dt} = \varepsilon_g \dot{c}_g \left(\bar{T}_s - T_{ge}\right) \quad (26)$$

with $c_s = M_s C_{ps}$, thermal capacity of the finite source in [J/K],

$\bar{T}_s$, average value of the soil temperature.

If the temperature $T_{ge}$ is a parameter (constant control variable), by integrating equation (26) one gets:

$$\frac{\bar{T}_s - T_{ge}}{\bar{T}_s(0) - T_{ge}} = e^{\frac{-t}{\tau_s}}$$

where $\tau_s = \frac{c_s}{\varepsilon_g \dot{c}_g}$ (or $\frac{c_s}{k_g A_g}$), thermal soil time constant.

The thermal soil time constant must be longer than the contact time at each exchanger (on the order of a minute) or the heat pump, so that the heat pump is always considered in steady state operation, while the source is imposing slow transient conditions:

$$\dot{Q}_g(t) = \varepsilon_g \dot{c}_g(\bar{T}_s - T_{ge})e^{\frac{-t}{\tau_s}} \quad (27)$$

It is therefore possible, using equation (27), to simulate the operation of the heat pump over a heating season for the various conditions discussed in section 2, and to analyze the sensitivity of this operation with respect to the parameters, in order to get an optimal sizing and a better control of the heat pump operation over time (operation with $T_S$ or $T_{ge}$, $\dot{c}_g, T_{conf}, \varepsilon_T$ as parameters, $\varepsilon_{évap}, \varepsilon_g$ as variables, and boundary conditions with $\dot{Q}_0, \dot{W}_0, or\ COP_0$ imposed).

### 3.2. *Some results of the Chambadal model of the geothermal heat pump*

An illustration of some results corresponding to the analytical expressions derived in section 2 for the cases of the Chambadal model relative to the geothermal heat pump are presented in figures 4-6 (the ordinate axe is given in arbitrary units).

For the case where the COP is imposed (section 2.1.4), figure 4 shows the evolution of the maximum of the $\dot{Q}_{g,arb}$ (dimensionless thermal power input from the geothermal source), as a function of the effectiveness of the geothermal source exchanger, for different values of the overall effectiveness $\varepsilon_T$. A strong influence of $\varepsilon_T$ on $\dot{Q}_{g,arb}$ variation is found, as well as effectiveness equipartition ($\varepsilon_g = \varepsilon_{evap} = 0.5$) for $\varepsilon_T = 1$, corresponding to $Max(\dot{Q}_{g,arb})$.

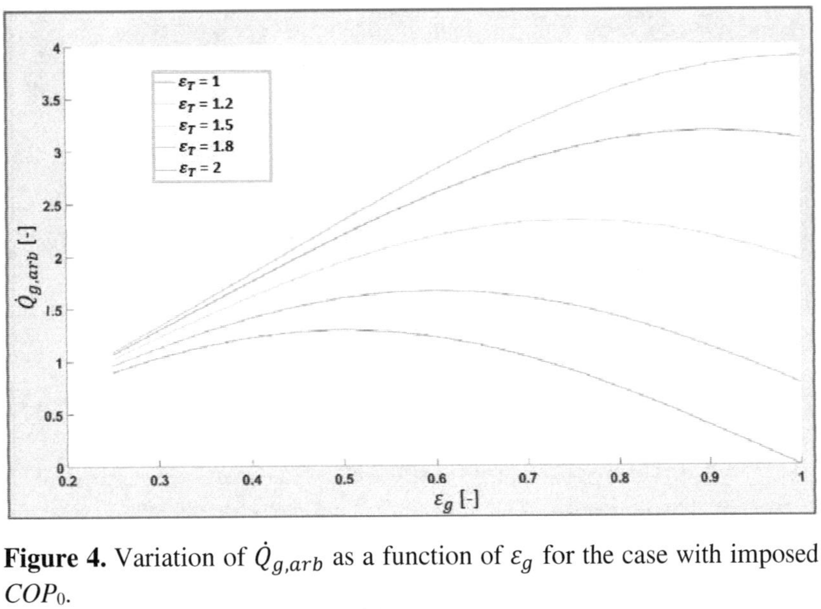


**Figure 4.** Variation of $\dot{Q}_{g,arb}$ as a function of $\varepsilon_g$ for the case with imposed $COP_0$.

In the case of an imposed $\dot{Q}_g$ at the cold source (section 2.1.2), figure 5 shows the existence of a minimum of the consumed mechanical power $\dot{W}$ as a function of $\varepsilon_T$. Despite of the fact that it is a very flat one, it proves that the sizing optimization through $\varepsilon_g$ can be achieved. Also, the curves in figure 5 show that it decreases and moves to higher values of the effectiveness $\varepsilon_g$ when the total effectiveness $\varepsilon_T$ increases.

The influence of $\varepsilon_T$ on the min($\dot{W}$) is small and it is confirmed through figure 6 where min($\dot{W}$) increases significantly with $\dot{Q}_0$, while the total effectiveness $\varepsilon_T$ makes the curves distinct only for high value of $\dot{Q}_0$.

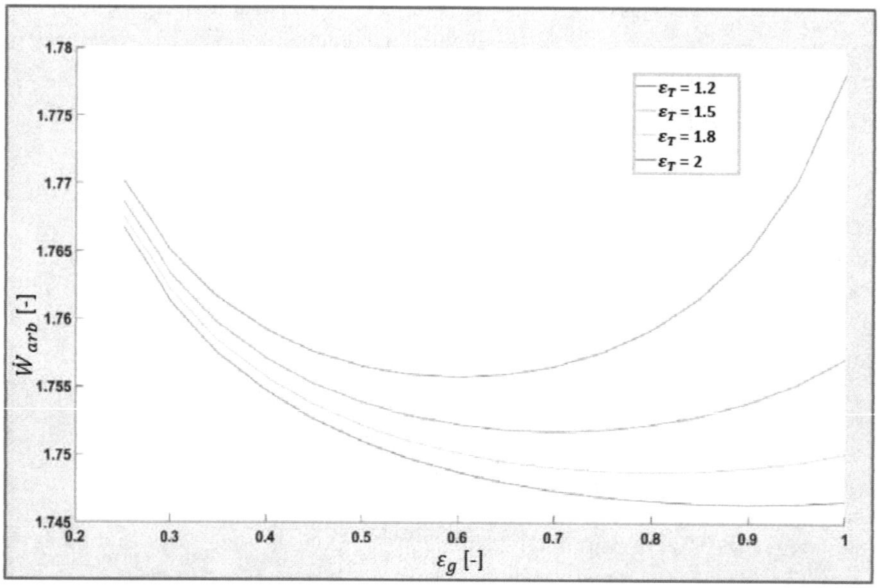


**Figure 5.** Variation of $\dot{W}_{arb}$ as a function of $\varepsilon_g$ for the case with imposed $\dot{Q}_0$.

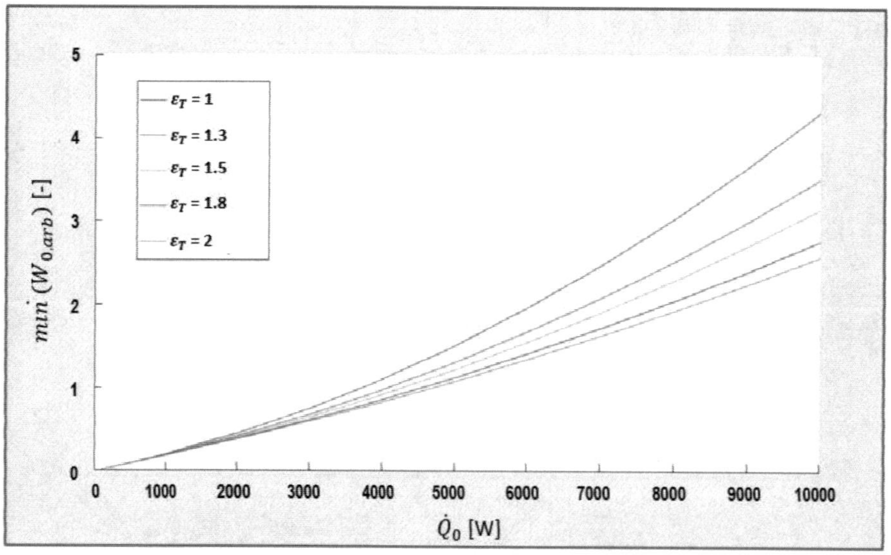


**Figure 6.** Variation of $min\ \dot{W}_{arb}$ as a function of $\dot{Q}_0$.

In the case of $\dot{W}_0$ imposed and for low values compared with the other heat flows, the following non-dimensional expression is obtained for the thermal power of the geothermal source:

$$\dot{q}_g \approx \frac{\dot{w}_0\, t_s}{1-t_s}\left[1 - \frac{\dot{w}_0}{1-t_s}\left(\frac{4}{\varepsilon_T} - 1\right)\right] \tag{28}$$

with $\dot{w}_0 = \frac{\dot{W}_0}{\dot{c}_g T_{conf}}$; $t_s = \frac{T_S}{T_{conf}}$.

IOP Conf. Series: Earth and Environmental Science **960** (2022) 012003 doi:10.1088/1755-1315/960/1/012003

The equation (28) becomes zero when $\dot{w}_0 = 0$ or $\dot{w}_0 = \frac{1-t_s}{\frac{4}{\varepsilon_T}-1} = \dot{w}_{lim}$.

It results that the maximum of $\dot{q}_g$ will be obtained for $\dot{w}_0 = \frac{\dot{w}_{lim}}{2}$, and will be given by:

$$Max(\dot{q}_g) = \frac{t_s}{2}\left(\frac{1}{\frac{4}{\varepsilon_T}-1}\right) \tag{29}$$

The maximum value of $\dot{Q}_g$ from equation (29) is an increasing function of ($T_S$, $\varepsilon_T$) from the point of view of operation, and then of ( $\dot{c}_g$, $T_{conf}$ ) from the point of view of heat exchangers dimensioning.

## 4. Conclusions and Perspectives

The paper explores the influence of the boundary conditions on the optimum operation and heat exchangers sizing of a geothermal heat pump. The reported elements complete those previously obtained in [1,4].

The modeling of the geothermal heat pump uses a Chambadal-type model well suited for the study and sensitivity analysis of the geothermal source boundaries conditions.

Original results are proposed and discussed. Thus, the existence of a minimum of mechanical power consumed is emphasized for imposed *COP* of the heat pump, respectively a maximum useful effect $\dot{Q}_g$ that is obtained for equipartition of the heat exchangers effectiveness.

In perspective, the extension of this approach to the Curzon-Ahlborn configuration of the heat pump is considered, as well as its validation by comparison with the existing results in the literature.

The extension of the model to the Curzon-Ahlborn configuration of the heat pump is currently developed, and then, its validation by comparison with existing results in the literature will be considered.

## References

[1] Gervaise C 2020 Entropic model of a heat pump *Master Thesis Lorraine University* (In French)
[2] Feidt M 2013 *Finite Physical Dimensions Optimal Thermodynamics* (Paris: Hermes-Lavoisier) (In French)
[3] Moyne C and Feidt M 2019 Low-temperature geothermal energy and heat pump *Revue générale du froid et du conditionnement de l'air*, March-April, 38-45 (In French)
[4] Pahud D and Hubbuch M 2007 Measurements and optimization of the installation with energy piles at the Dock Midfield at Zurich Airport *Final Report* (In French)

# Entropy analysis in spray cooling for dosing water injection

**S.O.L., Lacour[1], M. Feidt[2]**

[1]:University of Paris-Saclay, Laboratory INRAE-FRISE, Antony, France
[2]: University of Lorraine, Laboratory LEMTA, Nancy, France

E-mail: stephanie.lacour@inrae.fr

**Abstract.** Spraying water in air improves air-cooling capacity, which then relies on the evaporation of water. Even for small drop sizes, literature reports that the evaporation remains limited inside the spray and below saturation limits. In this article, we describe the evolution of thermodynamic quantities in a mixture of air and evaporating liquid water. A complete and coherent formulation is used to express enthalpy, entropy and chemical potential. At constant enthalpy, we show that the chemical equilibrium corresponds to an intermediate state in which droplet evaporation is not complete and entropy is maximum under certain conditions. Results are compared with some experimental values measured in a wind tunnel downstream of a spray. The calculated values are consistent with observations. Cooling efficiency is discussed for the various parameters, which are the amount of water, air temperature and ambient humidity. Then, the numerical approach is inverted in order to forecast the amount of water needed to reach a target cooling temperature. This numerical approach is used to set water flow depending on inlet flow conditions and cooling objectives.

## 1. Introduction

Misting cooling is a process consisting in spraying droplets of water into a flow of air to cool either air or water. Widely used to refresh surrounding air in summer, misting is also practiced in air treatment processes, for humidification purposes or to cool heating elements such as air conditioning condensers.

Sensitive to climatic conditions, effectiveness of misting has been widely studied, for example by Guan et al. in [1] to estimate its cooling potential in different regions of Australia. The work of Farmahini-Farahani et al. in [2] and [3] complete this analysis by describing exergy yields of mistcooling. Unlike Santos et al. in [4], which obtains exergy efficiencies close to 1, Farmahani-Farahani et al. report more moderate exergy efficiencies. The differences lie in definitions of exergy yields and their various formulations, which were analyzed in detail by Ratlamwala and Dincer in [5]. Indeed, the efficiency of evaporative processes, either energetic or exergetic, depends on whether it is based on saturation temperatures or enthalpies.

While the various parameters that influence evaporative yields have been studied, such as

air/water ratio in Lemouari et al. [6], droplet size and nozzle position in Tissot et al. [7], distance in Montazeri et al. [8], few theoretical elements allow a prior-test estimate about evaporative efficiency. However, experimental observations suggest that air/water mixture introduced in stream rapidely reach an equilibrium state from which evaporation becomes slow althoug vapour content is far from saturation ([7], [9], [10]). Numerical simulations confirm evaporative blockage, which results from mechanical, thermal and chemical balances occuring in air-water mixture ([8], [11]).
The exergetic approach offer a clear distinction between the 3 mechanisms involved during evaporation in mixtures to predict cooling efficiency. Sometimes, efficiency is also related to deviation from ideality or gradients within the mixture, affecting evaporation rate, like, Smrekar et al. in [12]. In this article, they use entropy and its topological distribution to optimize the cooling efficiency of a cooling tower. Mixing operations are also involved in chemical potential but are seldom used. Here, we assume the ideality of the mixture which corresponds well to a spray of small droplets. In this article, the detailed expressions of enthalpy, chemical potential and pressure are used to study the entropy of the mixture as well as mechanical, chemical and thermal equilibria. The chemical equilibrium coincides under certain conditions with the maximum entropy and is expressed from the operating conditions. It is shown that the chemical equilibrium agrees fairly well with experimental measurements. It is used to obtain an upper bound for cooling, related to partial evaporation, allowing to dose the corresponding sprayflow rate.

## 2. Model of mistcooling in wind tunnel

### *2.1. Process and hypothesis*

A mixture of air and liquid water undergoes conversion processes that involve thermal (cooling), mechanical (gas pressure, fragmentation and mixing of droplets), and chemical (evaporation) transfers between the phases and the components of mixture before reaching the equilibrium ([5],[13]). When mixing spray and air inside a wind-tunnel, transformations occurs in a space spatially constrained by the pipe's geometry: the transformations are therefore isochoric as long as the tunnel's section does not vary. The mixing is also supposed perfect: since the velocities of the phases are homogeneous, time and space are directly connected by the flow rate of air and water masses. We can therefore use a stationary or quasi-static approach to describe the thermo-chemical evolution of the mixture as it moves along the duct. The liquid-gas interface is then represented by the molar concentrations, which means a small size of the droplets (molar scale). Since the mix is not at equilibrium in its initial state, the evaporation of the liquid water occurs spontaneously. Chemical (evaporation), thermal (cooling)and mecanical (pressure) transfers in the environment are coupled.

### *2.2. Input data*

Humid air (ai) is introduced in wind-tunnel and liquid water (wl) is injected using nozzle spraying dropplets with a very small diameter. Air flowrate $V_g$ $(m^3/s)$, temperature $T_{ai}$ $(K)$, pressure $P_g$ $(Pa)$ and humidity $P_{wg}/P_g$ $(\%)$ are given, as well as water flowrate $m_{wl}$ $(kg/s)$ and temperature $T_{wi}$ (K). Molar flowrates (mol/s) are deduced from flowrates using molar mass of components ($\bar{M}_i$ $kg/mol$) and the total molar flowrate is constant:

$$n_t \quad = \quad \Sigma_i n_i = n_{as} + n_{wg} + n_{wl} = n_{as} + (n_{wg} + x) + (n_{wl} - x) \tag{1}$$

$n_g = n_{as} + n_{wg}$ refered to the gaseous molar rate (mol/s) introduced in the wind-tunnel and additionnal features about liquid-vapour transfer are used:

$$L_{vap} = h^0_{wg}(T_0) - h^0_{wl}(T_0) \quad C_{p,vap} = C_{p,wg}(T_0) - C_{p,wl}(T_0) \quad S_{vap} = S^0_{p,wg}(T_0) - S^0_{p,wl}(T_0)$$

$$C_{p,m} = \frac{n_{as}.C_{p,as} + n_{wg}.C_{p,wg} + n_{wl}.C_{p,wl}}{n_t} \qquad T_m = \frac{(n_{as}.C_{p,as} + n_{wg}.C_{p,wg}).T_{ai} + n_{wl}.C_{p,wl}.T_{wi}}{n_t C_{p,m}}$$

The fluid is made of two components (dry air $as$ and water $w$), water being in liquid $wl$ and vapour $wg$ phases. The Gibbs-Duhem relation states that the degree of freedom for the mixture made of n=2 components and p=2 phases is 2 at equilibrium. Thus, state variables are written using 2 variables $T_m$ (mixture temperature) and $x$ molar rate of evaporated water (mol/s). After fragmentation of liquid, water droplets are mixed into the gaseous flow to obtain an assumed perfect and homogeneous mxiture. Then, evaporation starts and $x$ moles of water changes from liquid to vapour phase while temperature decreases. Air and water masses are conserved but water amount in each phase varies. Thermodynamic functions related to evaporation in the mixture are given in the following.

### 2.3. *Enthalpy, chemical potential and entropy of mixture*

*2.3.1. Enthalpy* The system is described with terms related to mechanical (pressure), thermal and chemical equilibrium. Fluid internal energy $dU$ is $-P.dV + T.dS + (\mu_{wg} - \mu_{wl}).dx$ whereas its enthalpy $dH$ is $V.dP + T.dS + (\mu_{wg} - \mu_{wl}).dx$ ([14]). Each of these terms are detailled in Table 1. Enthalpy is the sum of elemental enthalpies weighted by molar rates and its differential form is stated with reference to a standart state ([14]). This latest is defined by $P_0 = 101\,325$ Pa and $T_0 = 298$ K, with humid air containing $n_{wg}$ moles of water vapour coexisting with $n_{wl}$ moles of liquid water. Formation enthalpies, entropies and calorific values are given by thermodynamic tables ([15]). Calorific values are supposed constant and no thermal gradient exists between phases when temperature changes from mixing temperature down to final equilibrium temperature.

Table 1: Enthalpy, pressure, entropy and chemical potential equations (W)

| | $dH$ | $V.dP$ | $T.[dS]$ | $(\mu_{wg} - \mu_{wl}).dx$ |
|---|---|---|---|---|
| Chemical | $(L_{vap} + C_{pvap}.\Delta T).dx$ | | $T.\left[S_{vap} + C_{pvap}.\log \frac{T}{T_0}\right].dx$ | $(L_{vap} + C_{pvap}.\Delta T)$ -T.$\left(S_{vap} + C_{pvap}.\log \frac{T}{T_0}\right)$ |
| | | $RT.dx$ | $T.[-R.dx]$ | |
| Mechanical | | | $-T\left[R \log \frac{n_t}{n_l - x}.dx\right]$ | $+RT.\log \frac{n_t}{n_l - x}$ |
| | | | $-T\left[R \log \frac{n_{vap}+x}{n_g+x}.dx\right]$ | $+RT.\log \frac{n_{vap}+x}{n_g+x}$ |
| | | | $-T\left[R \log \frac{P}{P_0}.dx\right]$ | $+RT.\log \frac{P}{P_0}$ |
| Thermal | $(n_t.C_{p,m} + x.C_{p,vap}).dT$ | | $T.[(n_t.C_{p,m} + x.C_{p,vap}).dT/T]$ | |
| | | $R(n_g + x).dT$ | $-T.[(n_g + x).R.dT/T]$ | |
| Other | | | $-T.[\mu_0 dx/T]$ | $\mu_0$ |

*2.3.2. Pressure* The ideal gas law is used to described pressure ($P = (n_g + x).R.T/V_g$) where $R\ (J/(mol.K)$ is the ideal gas constant and $V_{ai}$ the constant volume (rate $m^3/s$) of tunnel. Hence, the impact of liquid on pressure is neglected. As initial condition differs from the standart state, flowing work (W) is needed to transfer fluids (mainly air), which is written at standart pressure:

$$\delta W \quad = \quad -P_O.(V - V_0) = n_g.R.(T_m - T_0) \tag{2}$$

This open system is isenthalpic, so $dU$ is $-V.dP - PdV$. Therefore, pressure change related to cooling $V.dP$ are added to flowing work $\delta W$ into internal fluid energy.

*2.3.3. Chemical potential* Some thermochemistry is reminded here in order to link our formulae with usual expressions used for moist air. Like enthalpy, chemical potential is defined up to a constant. For the reactive pure components (liquid and vapour water), it is related to entropy and enthalpy by:

$$\mu_i^{pure} \quad = \quad h_i(T) - T.S_i(T) = h_i^0(T_0) - T.S_i^0(T_0) + C_{p,i}.\left[T - T_0 - T.\log(\frac{T}{T_0})\right] \tag{3}$$

In mixture, component concentrations modify potential, what is written for an ideal gas (vapour) and a very dilute ideal solution (droplets):

$$\begin{aligned} \mu_{wg}(T, P, c_{wg}) \quad &= \quad \mu_{wg}^{pure}(T) + R.T \log \frac{P}{P_0} + R.T \log \frac{P_{wg}}{P} \\ \mu_{wl}(T, P, c_{wl}) \quad &= \quad \mu_{wl}^{pure}(T) + R.T \log c_{wl} \end{aligned}$$

Change of chemical potential in a reactive mixture is generally written $\Delta_r\mu = \Delta_r\mu^0 + R.T \log Q_r$ which is also $RT.log(K(T)/Q_r)$. Indeed, the equilibrium constant $K(T)$ is more often used than $\Delta_r\mu^0(T)$. K(T) depends on temperature, according to a relation written with $log(K) = A - B/T$. By comparing these 2 expressions, parameters of equilibrium constant are expressed by $A = S_{vap}/R$ and $B = L_{vap}/R$. Hence, entropy and enthalpy of formation are related to equilibrium constant for the evaporation process.

Concentrations are related to the reactive quotient $Q_r$, which rates product over reactant concentrations in monophasic. Here, dilution ratio is needed for liquid water and partial pressure for water vapour. With only moist air, $Q_r = P_{wg}/P_O$ is egal to equilibrium constant $K(T)$, which leads directly to the saturation law for moist air if $P_{wg} = P_{sat}$.

In diphasic mixture, liquid water impacts dilution ratio of ideal solution $RT.\log((n_{wl} - x)/n_t)$ whereas air humidity is involved in partial pressure for vapour $RT.\log((n_{wg} + x)/(n_g + x))$. These both terms modify chemical equilibrium. Thermochemistry analysis reveals that adding liquid water improves evaporation whereas adding water vapour reduces it. Dry air does not change equilibrium position.

Chemical potential is defined up to a constant which is assessed while observing that no evaporation occurs if initial mixture temperature equals air dew point temperature $T_{dp}$). The chemical potential is null at x=0. Then, the first two lines in Table 1) are equal to $R.T_{dp}.log(n_{wg}/n_g)$, which is air saturation. This is injected in the potential equation to assess $\mu_0$, which sizes in fact the desequilibrium involved by mixing.

$$\mu_0 = RT_m.\log \frac{n_{wl}}{n_t}.\frac{T_0}{T_m} \tag{4}$$

Chemical potential in Table 1 implicitely contains saturation law for moist air through entropy and enthalpy formation. The main change of our expression, compared to usual expressions in exergetic litterature (e.g. [16],[17],[13]) lies in introducing concentration for liquid water. Thermochemistry analysis indicates that this term has an impact on the local chemical equilibrium inside wind tunnel but it is cancelled in exergy. This is because the equilibrium is stated with respect to the environment (infinite air volume), where a complete evaporation up to air saturation is reached.

*2.3.4. Entropy* Like enthalpy, mixture entropy is the sum of component entropies, to whom a production term is added. The production term includes diffusion terms (liquid dilution and partial pressure changes) included in the chemical potential. Thus, such an entropic expression, found in Table 1, make up for pressure $V.dP$ and chemical variations, as needed to cancel $V.dP + T.dS + (\mu_g - \mu_l).dx$ and get a constant enthalpy.

## 3. Isenthalpic vaporisation and chemical equilibrium

### *3.1. Evaporation at constant enthalpy*

Evaporative cooling is a chemical conversion at constant enthalpy, which is described by cancelling the first column of Table 1.

$$\frac{dT}{L_{vap} + C_{p,vap}.(T - T_O)} = \frac{-dx}{n_m.C_{p,m} + x.C_{pvap}} \tag{5}$$

The differential is integrated between initial mixing without evaporation and local conditions given by $[T_m, T]$ for temperatures and $[0, x]$ for moles of evaporated liquid. This gives:

$$T = \frac{n_m.C_{p,m}.T_m - x.(L_{vap} - C_{p,vap}.T_0)}{n_m.C_{p,m} + x.C_{p,vap}} \approx T_m - x.\frac{L_{vap} + C_{pvap}(T_m - T_0)}{n_m.C_{p,m}} \tag{6}$$

Temperature drop is linked to endothermic reaction for vaporisation. At the first order, the relation is approximed by the line whose slope is $\beta = (L_{vap} + C_{pvap}(T_m - T_0))/(n_t.C_{p,m})$. Equation 6 represented on Figure 1, with circles for the whole expression and by the line for the approximation. The imperceptible difference ascertains that the linear approximation can be further used.

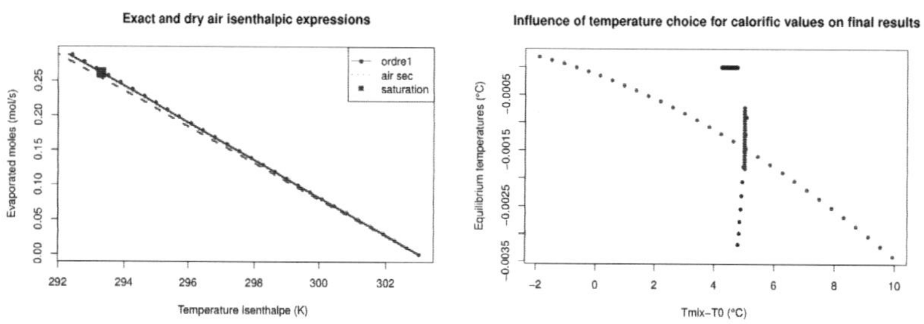


Figure 1: On the left, temperature according to liquid evaporation moles using present- and air- enthalpy approximation and on the right, equilibrium temperature gradient using calorific values assessed at $T_{air}$ or $T_0$.

### *3.2. Chemical equilibrium*

Table 1 ensures cancelling terms for enthalpy, when assessed from entropy, pressure and chemical potential. It also emphasizes that chemical equilibrium corresponds to entropy maximum only if pressure terms are cancelled. Equilibrium equation is reached when $\mu_g - \mu_l$ equals zero, which is written in Equation 7 using only $x$, amount of liquid moles that are evaporated. For given initial conditions, equation root is numerically solved using R software ([18]). Root of Equation 7 expresses the amount of evaporated water at chemical equilibrium and is used to derive other quantities, like temperature with Equation 6.

$$\begin{aligned} 0 = & \left[S_{vap} + C_{pvap}.\log\frac{T_m - \beta.x}{T_0}\right] - \beta.\frac{n_t * C_{p,m} + x.C_{p,vap}}{T_m - \beta.x} \\ & - R\log\left(\frac{n_{wg} + x}{n_g}\frac{n_t}{n_{wl} - x}\frac{T_m - \beta.x}{T_0}\right) - R\frac{T_m}{T_m - \beta.x}\log\left(\frac{n_{wl}}{n_t}\frac{T_0}{T_m}\right) \end{aligned} \tag{7}$$

IOP Conf. Series: Earth and Environmental Science **960** (2022) 012004 doi:10.1088/1755-1315/960/1/012004

### *3.3. Setting water injection*

The aim is to cool air down to target temperature, $T_{target}$ which is necessarily between mixing and adiabatic saturation temperatures. Setting water with this objective means that target temperature equals equilibrium temperature. Molar conservation implies $n_t = n_g + n_{wl}$, where liquid water moles $(n_{wl}\ (mol/s)$ is unkown. Assuming same temperature $T_i$ for air and water at inlet, the isenthalpic equation expresses now the relation between moles to evaporated and physical mixture properties:

$$x \quad = \quad -(T_{target} - T_i).\frac{L_{vap} + C_{p,vap}.(T_i - T_0)}{n_g.C_{p,g} + n_{wl}.C_{p,wl}} \tag{8}$$

Combined with equilibrium equation 7, the following equation is obtained and depends only on injected liquid moles:

$$\begin{aligned} 0 \quad &= \quad S_{vap} + (C_{p,vap} - R).\log\frac{T_{target}}{T_0} - \frac{L_{vap} + C_{p,vap}.(T_i - T_0)}{T_{target}} + C_{p,vap}.\frac{T_{target} - T_i}{T_{target}} \\ &- \quad R.\log\left(\frac{n_{wg} + \alpha * (n_g.C_{p,g} + n_{wl}.C_{p,wl}}{n_{wl} - \alpha.(n_g.C_{p,g} + n_{wl}.C_{p,wl})}.\frac{n_g + n_{wl}}{n_g}\right) - R.\frac{T_i}{T_{target}}.\log\left(\frac{n_{wl}}{n_g + n_{wl}}\frac{T_0}{T_{target}}\right) \end{aligned} \tag{9}$$

Where $\alpha = -(T_{target} - T_i)/(L_{vap} + C_{p,vap}.(T_i - T_0)$. Here again, an equation solver is used to extract root of equation 9 defining the amount of water needed to get sufficient evaporation at chemical equilibrium for cooling air down to target temperature.

## 4. Results

### *4.1. Reference case*

A reference case, corresponding to detailled inlet conditions in Table 2 is used for sensitivity analysis. These conditions are used to study the impact of temperature used in the assessement of calorific values (see Figure 1 on right). It shows that final equilibrium temperature weakly changes with the various $C_p$ values.

Table 2: Input data in the reference case

| $T_{ai}$ | $HR_a$ | $P_{ref}$ | $V_{gaz}$ | $m_{wl}$ | $T_{wl}$ |
|---|---|---|---|---|---|
| °C | (%) | Pa | $m^3/s$ | kg/s | °C |
| 30 | 40 | 101 325 | 1 | 0.004 | 20 |

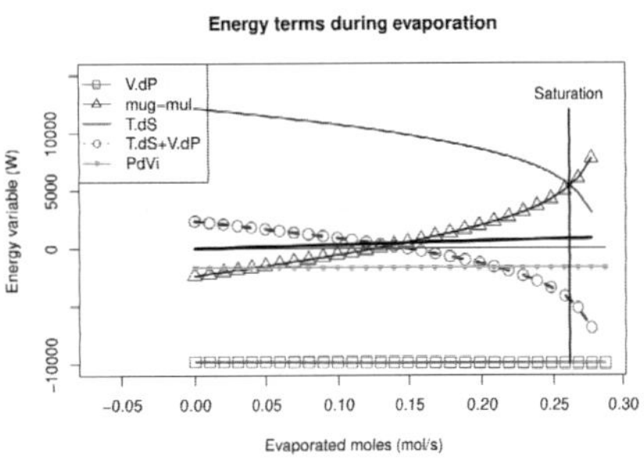


Figure 2: Energy variations along the evaporation process

### *4.2. Impact of evaporation on energy variables*

Energy terms in Table 1 are computed for the reference case with varying the amount $x$ of evaporated moles, which is represented on Figure 2. The orange line represents flowing work for fluid, which is constant over evaporation process. The green line, used for the pressure term, is much more negative than flowing work but stays also quite constant over evaporation process: mixture pressure decreases with temperature in the wind-tunnel and its derivative is quite constante, showing the small effect of molar changes in gas. Chemical potential, in blue, is negative at starts, as needed for spontaneous reaction, but becomes zero in the middle, blocking there the evaporation reaction. Chemical potential curve intersects entropy curve in red at adiabatic saturation: there, these two terms balances pressure variation. But this state can not be achieved spontaneously as evaporation is blocked before. The black curve is the sum of pressure, entropy and chemical terms, which is enthalpy and stays near zero along the x-axis. The small slope is attributed to the first order approximation. Entropy derivative is positive at starts, showing mixture entropy increase when beginning evaporation. If pressure terms are removed from entropy expression, curve with red circle is obtained and then chemical equilibrium matches pseudo-entropy maximum. If entropy derivative is not zero, evaporative cooling can not be adiabatic unless pressure derivative is zero. Or pressure contribution in entropy derivative represent mechanical irrerversibilities that should not be included in heat related to entropy. Such irreversibilities could also explain pressure drop and additionnal work needed at fan in evaporative systems.

### *4.3. Studying the chemical equilibrium*

Chemical equilibrium is reached before adiabatic saturation in our reference case. As equilibrium depends on mixing conditions, Figure 3 shows equilibrium positions in the psychrometric chart for various operating conditions.

The green doted line is for inlet air temperature variations, the red doted one is obtained by changing air humidity at inlet and the blue doted one by changing liquid flowrate. Solide lines represent the related evolution of the equilibrium. Cross of the three curves is the solution for the reference case. While increasing air temperature, evaporation rate is increased as showed by the humidity ratio of the green curve. If air humidity is increased, the cooling effect is reduced since the equilibrium temperature rises on the red curve. Increasing water flowrate improves cooling on the blue curve close to the isenthalpic curve.

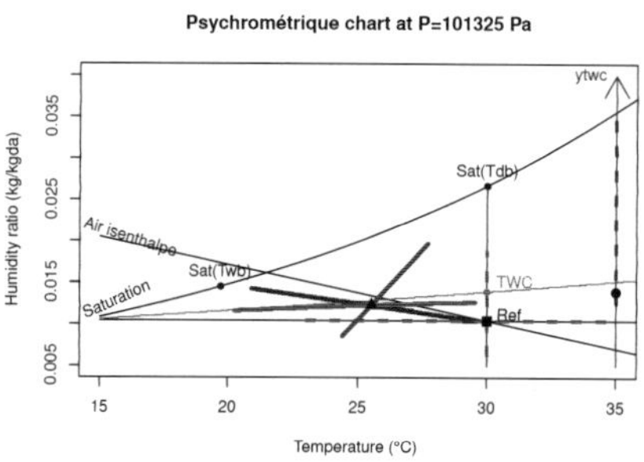


Figure 3: Chemical equilibrium in psychrometric chart for various inlet conditions

The gap between air isenthalpic and blue curves arises from water temperature: the more water, the more precooling effect of mixing. This shifts the real origin of isenthalp (mixing temperature) on left. A graphic method is also proposed here for approaching chemical equilibrium: the violet exponential curve, passing through $[T_{dp}, y_{ai} = n_{wg}.\bar{M_{wg}}/(m_{as}.\bar{M_{as}})$ and $[T_{ai}, y_{twc} = y_{ai}+n_{wl}.\bar{M_{wl}}/(m_{as}.\bar{M_{as}})]$ intersects air isenthalpic curve close to equilibrium solution.

### 4.4. Comparison between modelled and observed data

On Figure 4, measurements of spraying in wind-tunnel found in [7] are reported. Green and blue marks are used for the two positions of measurements, made for several inlet conditions. These pairs of measurements reveals evaporation blocking. In red are reported calculated chemical equilibrium by solving Equation 7, that is independant of positions or time. Red triangles are computed assuming standart atmospheric pressure in the wind tunnel because pressure measurements are not specified in the paper. Computed solutions always overestimates cooling in almost all conditions: our estimate is a majorant of cooling efficiency.

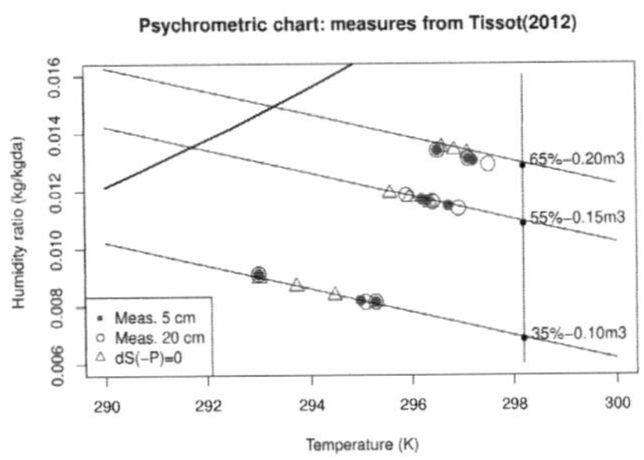


Figure 4: Comparison of calculated chemical equilibrium and measurements published in Tissot(2012)

However, evaporated fluxes matches quite well measurements and calculation gives a correct magnitude for cooling efficency according to operating changes. A lot of imperfections (droplet size, non ideal mixing) could explain why the chemical optimum is not completely achieved in experiments.

### 4.5. Setting water injection according to cooling objectives

Equation 7 sizes a correct majorant for cooling air with spray in wind tunnels for various inlet conditions. So the inverted Equation 9 may be used to size minimal spray flowrate needed to ensure a target temperature in cooling. Figure 5 shows the cooling effect obtained at chemical equilibrium for various water flowrates (reference case for other variables).

Cooling, shown with the black curve, is increased when rising spray flowrate. If the flowrate exactly matches vapour needs to obtain adiabatic saturation ($m_{wl,sat}(T_{wb})$), cooling is around 5°C when cooling potential (orange curve) is above 10°C. Evaporation always stays uncomplete and excess water is needed to increase cooling. But even if equilibrium temperature is lowered by increasing spray flowarte, evaporation rate (red curve) decreases. The more water injected, the more cooling but also the more unused water.

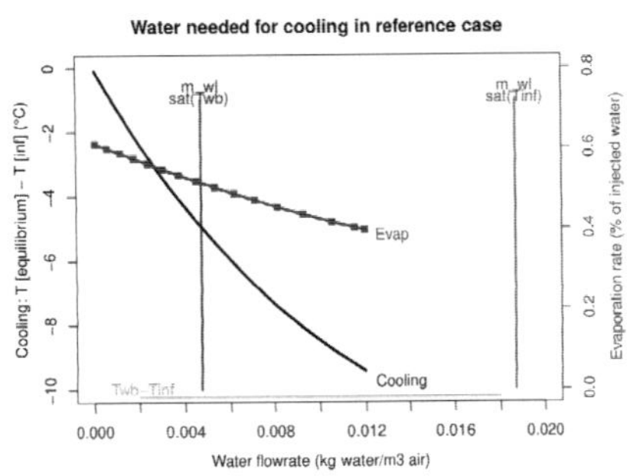


Figure 5: Cooling reached at chemical equilibrium according to injected liquid water flowrate

Cooling slowly converges near adiabatic saturation when water flowrate tends to saturation flowrate at air temperature $m_{wl,sat}(T_{air})$.

## 5. Conclusion

This article aims to study thermodynamic variables in misted air used for evaporative cooling. Expressions of enthalpy, chemical potential and entropy are detailled for an ideal mixture of air and liquid water in an isochoric conversion. Analysis of the mixture chemical potential shows that evaporation of droplets occurs spontaneously in air. But evaporation reaction ends before reaching adiabatic saturation, in a state which corresponds in fact to a local chemical equilibrium. This chemical equilibrium varies as a function of operating conditions and a graphical method is proposed to approximate this position in the psychrometric diagram. Position of chemical equilibrium is found to be a reliable upper bound, compared to experimental measurements about spraying in wind-tunnels. Uncertainties on spray cooling efficiency does not lie in evaluating the physical properties of fluids but rather in mixing imperfections. The thermochemical analysis thus offers a method for dosing the water to inject in order to cool air while accounting for evaporative blockage. Results also show that the evaporative cooling is either isenthalpic or adiabatic, but not both, except by considering that mechanical irreversibilities associated with the pressure terms in the entropy are not integrated in heat. Further work is needed to study adiabaticity of evaporative cooling and requires additional measurements of pressure.

## Acknowledgments
Authors wish to thank Jean-Marie Alessandrini et Jacques Ribéron, from french Scientific and Technical Center for Buildings, for sharing documents and guiding work onto applied perspectives.

## References
[1] Guan L, Bennett M and Bell J 2015 Energy and Buildings **108** 185–194
[2] Farmahini-Farahani M and Pasdarshahri H 2011 Journal of Mechanical Research and Applications **3**
[3] Farmahini-Farahani M, Delfani S and Esmaeelian J 2011 Journal of Mechanical Research and Applications **3** 185–194
[4] Santos J, Barros G, Gurgel J and Marcondes F 2013 International Journal of Refrigeration **36**
[5] Ratlamwala T and Dincer I 2013 International Journal of Refrigeration **36** 1142–1153
[6] Lemouari M and Boumaza M 2010 International Journal of Thermal Sciences **49** 2049–2056
[7] Tissot J, Boulet P, Labergue A, Castanet G, Trinquet F and Fournaison L 2012 International Journal of Thermal Sciences **60** 23–31
[8] Montazeri H, Blocken B and Hensen J L M 2015 Building and Environment **83** 129–141
[9] Zhang F, Bock J, Jacobi A M and Wu H 2014 APPLIED THERMAL ENGINEERING **63** 528–540
[10] Hamlin S, Hunt R and Tassou S A 1998 Applied Thermal Engineering **18** 1139–1148
[11] Raoult F, Lacour S, Carissimo B, Trinquet F, Delahaye A and Fournaison L 2019 Applied Thermal Engineering **149** 960–974
[12] Smrekar J, Oman J and Širok B 2006 Energy Conversion and Management **47** 1086–1100
[13] Qureshi B and Zubair S 2006 International Journal of Thermal Sciences **46** 188–198
[14] Borel L and Favrat D 2005 Thermodynamique et énergétique (Thermodynamique et énergétique no vol. 1) (Presses polytechniques et universitaires romandes) ISBN 9782880745455 URL https://books.google.fr/books?id=ApHgAgC81G0C
[15] Burcat A and Ruscic B 2005 Third Millennium Ideal Gas and Condensed Phase Thermochemical Database for Combustion with Updates from Active Thermochemical Tables (TAE 960 vol ANL-05/20) (Argonne National Laboratory)
[16] Bejan A 2006 Advanced Engineering Thermodynamics, Third Edition (John Wiley & Sons) ISBN 0471677639
[17] Chengqin R, Nianping L and Guangfa T 2002 Building and Environment **37** 1045–1055
[18] R Core Team 2015 R: A Language and Environment for Statistical Computing R Foundation for Statistical Computing Vienna, Austria URL https://www.R-project.org/

IOP Conf. Series: Earth and Environmental Science **960** (2022) 012005 doi:10.1088/1755-1315/960/1/012005

# Innovative energy efficient thermal insulation to ensure indoor comfort in nZEB buildings

**Emilia-Cerna Mladin[1], Madalina Calbureanu[2,*], Raluca Malciu[2] and Diana Calbureanu[2]**

[1] University POLITEHNICA of Bucharest, Mechanical and Mecathronic Engineering Department, 313 Splaiul Independentei, 060042, Romania

[2] Faculty of Mechanics, University of Craiova, 13 A.I. Cuza, Craiova, 200585, Romania

E-mail: madalina.calbureanu@gmail.com

ORCID : https://orcid.org/0000-0002-2557-4460

**Abstract**. The paper presents the analysis of water and heat transfer through walls being thermally insulated with vegetal materials. The analysis identifies the risk for condense accumulation in the outer layers of external walls of a common residential building located in the coldest climatic regions of Romania. Different wall structures and insulation thicknesses are systematically considered with statistical extreme temperature and humidity outdoor values. Results are useful in designing nZEB individual houses with green and sustainable technologies that also provide energy savings and indoor conditions for good comfort and health in these desired building concept.

## 1. Introduction

In the current context of designing energy efficient buildings that provide occupants with a comfortable and healthy interior environment, it is very important that during major renovations or new buildings design to use innovative materials that are considered green/ecological with very good thermal insulating properties and with sufficient permeability to water vapor transfer. Considering the predominant volume of the existing built fund, it is necessary to rigorously analyze especially the buildings in climatic areas with extreme conditions, in order to identify energy efficient and non-polluting solutions, but also available and accessible on the Romanian market. These solutions must also meet the sustainability requirements of a building, especially since the value of a building is estimated taking into account the costs throughout the life cycle (including repair and maintenance costs which represent about 40% of total costs). The location of the building must also be considered as well as orientation, vicinity and access to utilities. Only taking into account all these parameters, sustainable solutions can be chosen and the effects can be correctly predicted for a long enough period to motivate the necessary investment [1].

From the multitude of ecological materials used more or less in recent years (wool, mineral wool, cellulose etc.), there is a special interest in the use of cork, a renewable plant material produced naturally in certain areas of the planet. For this type of material, the permeability to water vapor was studied, in order to reduce the condensation phenomenon, corrosion and resistance to water or to the polluting elements in the environment of the construction elements, as well as the durability of the thermal insulation properties. Last but not least, it is important that the implementation is easy and therefore inexpensive. Following studies to date, the use of cork compared to the commonly used solution based on expanded polystyrene (EPS) has led to the conclusion that there is no significant difference in thermal insulation performance, but CO2 emissions associated with the manufacture and installation of EPS are double those associated with the use of cork [2]. From cork, thermal insulation panels can be made for all building's envelope elements (attic floor, terrace floor, exterior walls, floor tile, floor with built-in radiant heating). An additional advantage is that the cork agglomerate decreases the sound level due to the absorption effect which reduces the reverberation times. Thus, a thermally insulated plate with cork shows a reduction of the transmission of vibrations and noise caused by possible impacts; for this, it is important to maintain the discontinuity between the floor screed and the surrounding walls, thus eliminating marginal transmissions [3].

All existing studies have shown that the material is economically efficient in terms of initial investment and longer-term profitability. In short, the advantages of using this type of insulating material, highlighted in studies conducted so far [4], are:

- 100% natural and with a negative carbon footprint;
- thermal and acoustic insulator, including anti-vibration, with high dimensional stability;
- withstands temperatures between -180 C and +120 C, ensuring a good quality of indoor air, and in case of fire does not emit toxic gases;
- has good water vapor permeability, but is impermeable to liquids;
- cannot be attacked by rodents;
- it is resistant to compression (good mechanical properties);
- it is 100% recyclable and reusable in other applications etc.

## 2. Comparative study. The advantages of using cork.

Widely used insulating materials have different mechanical and physical properties. The choice of the optimal solution depends on major properties, such as:

- capacity for mechanical resistance to compression and durability;
- the degree of difficulty with which it can be put into operation and the fire behavior;
- thermal conductivity (W/m·K) and water vapor permeability.

The thermal insulation material used for this study is cork, a recyclable material that comes from the bark of the cork oak (Quercus Suberus), the cork being harvested once every 25 years (figure 1).

**Figure 1.** Cork shaft bark. Source: Ecoesmas [12].

Cork is sold in various forms: panels, sheets, blocks, granules, crushed, etc. Of all these, the most interesting are the expanded cork panels (black) and the agglomerated cork panels (white); the first one has the cork glued with the help of its own resin, and in the case of the white ones, recyclable cork is used and it is glued with food polyurethane adhesive. Particleboard panels are better from an ecological point of view because they are recycled/recyclable materials and require much lower temperatures for their manufacture.

Being an antibacterial material (does not allow the development of fungi, mold) and rot - proof, the cork is ideal for use in areas with possible high moisture concentrations, such as basements and underfloor. The panels are rigid, which makes their installation easy and practical.

In order to highlight the advantages of using cork, a comparative table between cork and basalt mineral wool is presented below (table 1).

**Table 1.** Table comparing properties of cork/basalt mineral wool.

| Name | *Basalt mineral wool* | *Cork (ICB)* |
|---|---|---|
| Origin | Mineral | Bio |
| Conductivity λ (W/m.K) | 0,03 – 0,05 | 0,034 – 0,100 |
| Resistance factor to water vapor diffusion (μ) | 1 | 5 - 30 |
| Inflammable | NO | NO |
| Density (kg/m³) | 15-80 | 55-60 |
| Thickness (K=0,2cm) | 17,50-25 | 22,50-27,50 |
| Approximate price €/m² | <5 | <25 |
| Presentation | Panels, mattresses, rollers. The dimensions of the panels can be 50x100cm. These products can be delivered with or without casing made of aluminum foil or craft paper. | Panels and rollers. |
| Protection measures during installation | Eyes, respiratory system (risk of asthma) and skin; may cause dermatological conditions. | NO<br>However, cork dust could be a problem, it is better to avoid inhalation. |
| Energy cost [1] | 15 - 25 | 1 - 25 |
| Recyclable product content (0-3)[2] | 1 | 0 |
| Biodegradable | NO | YES |
| Ecological aspects | The elaboration of the raw material implies a high consumption of energy and ecological damages, but also on the health. | The elaboration of the raw material implies a high consumption of energy and ecological and health damages. |
| Constructive properties | A very good capacity as thermal and acoustic insulation. Low capacity to accumulate energy. | A good capacity as thermal and acoustic insulation. |
| | Used for thermal insulation of interior and exterior walls, ventilated facades, floors, terraces, roofs, thermal systems frame. | Used in thermal insulation of floors and walls. |
| Use and real life time | Stable in the face of moisture. Elastic. Resistant to use. Stable in shape. There is no danger of sheltering insects. Not recommended for finishes, as it accumulates dirt. Long life. | Stable in the face of moisture. Undoubtedly, if the exposure is long lasting, moisture or insects can appear. Not recommended for finishes, as it accumulates dirt. He has a long life. |

Energy consumption and CO2 emissions associated with the manufacture of panels are different, being lower in the case of cellulose and higher in other types of thermal insulation commonly used.

Figure 2 shows a comparison of the energy used, emissions and the level of toxicity of cellulose to expanded polystyrene and mineral wool, in the process of manufacturing them.

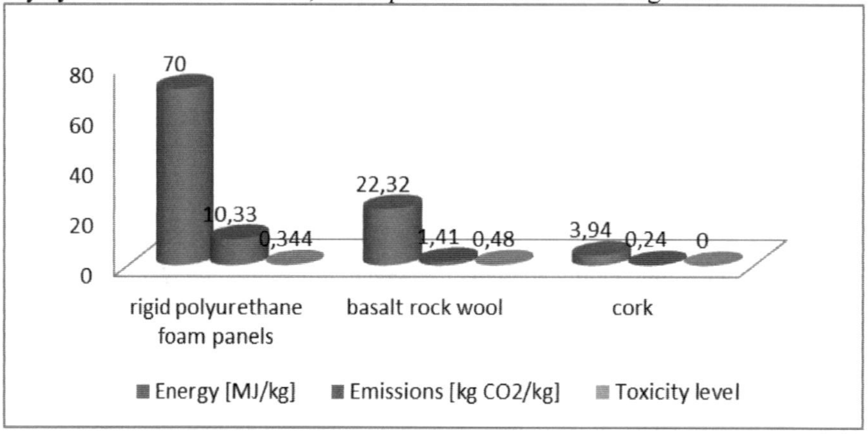


**Figure 2.** Energy, CO2 emissions and toxicity levels for cork, polyurethane and mineral wool in their manufacturing process.

It can be easily noticed that upon all three criteria, cork is superior to other types of classic thermal insulation.

From the point of view of use, cork can be used for thermal insulation of house walls on wooden or brick structure, in the case of attic, roof, thermal insulation under parquet, sound insulation between levels, etc. The products have no smell, have a pleasant appearance, and their assembly is very simple. Figure 3 shows the value of thermal resistance of cork compared to other conventional thermal insulators.

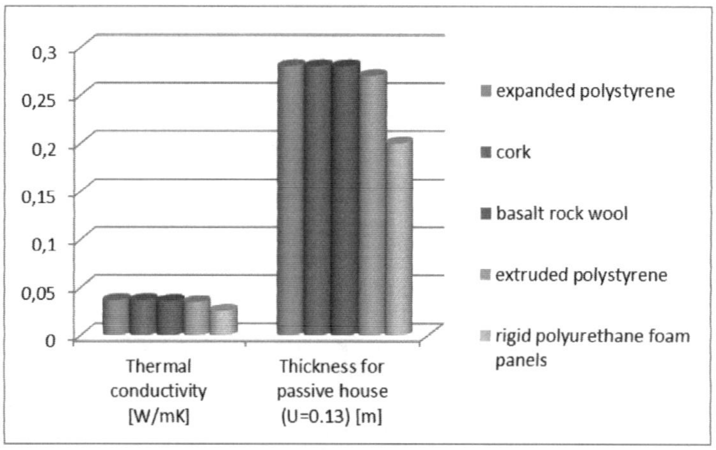


**Figure 3.** The thermal conductivity of cork compared to other classic thermal insulators and the thickness of the layer required for the passive house (U=0,13 W/m$^2$K).

## 3. Mass transfer by thermal insulation from cork

A residential house from climate zone IV with indoor and outdoor climatic parameters from table 2 is used as a study model and the risk of condensation and water accumulation in the cold season is evaluated for this [5-7].

It is considered an indoor air temperature of 20°C, and the average annual outdoor air temperature is 6.50°C. The composition and characteristics of the construction element are presented in table 3.

**Table 2.** Indoor and outdoor climatic parameters necessary to verify the progressive non-accumulation of water.

| Indoor climatic parameters | | |
|---|---|---|
| Indoor air temperature | 20 | °C |
| Relative humidity of indoor air | 60 | % |
| Heat transfer coefficient at the inner surface | 8 | W/m$^2$°C |
| **External climatic parameters** | | |
| Average annual temperature | 6.50 | °C |
| Statistical minimum temperature | -21 | °C |
| Relative humidity of the outside air | 80 | % |
| Heat transfer coefficient at the outer surface | 24.00 | W/m$^2$°C |

**Table 3.** The composition and characteristics of the construction element.

| Code | The layers of the building element (they are mandatory to be inserted from the inside out) | Layer thickness (m) | Thermal conductivity (W/m°C) | The vapor permeability resistance factor (−) |
|---|---|---|---|---|
| 52 | Mortar of lime | 0.0100 | 0.700 | 5.30 |
| 202 | Brick masonry with vertical holes, GVP type | 0.3500 | 0.580 | 4.50 |
| 206 | Cork | 0.1000 | 0.034 | 5.00 |
| 51 | Cement mortar and lime | 0.0100 | 0.870 | 8.50 |

A wall made of GVP (vertical hollow brick) with a thickness of 35 cm was considered, and the thickness of the thermal insulation is 10 cm, similar to that of a classic thermal insulation with Basalt mineral wool [8-11]. The outside air temperature at which the condense appears is –10.78 °C, at a relative humidity of 85%.

**Table 4.** Temperature and partial pressures on the surfaces of the layers.

| Indoor climatic parameters | | | | |
|---|---|---|---|---|
| Indoor air temperature | | | 20 | °C |
| Relative humidity of indoor air | | | 60 | % |
| Heat transfer coefficient at the inner surface | | | 8 | W/m$^2$°C |
| **External climatic parameters** | | | | |
| The outside air temperature at which condensation occurs | | | -10.78 | °C |
| Relative humidity of outdoor air | | | 85 | % |
| Coefficient of thermal transfer of the outer surface | | | 24 | W/m$^2$°C |
| **Crt. No.** | **Layers surfaces** | **Partial pressure** (Pa) | **Saturation Pressure** (Pa) | **Temperature** (°C) |
| | Indoor air | 1402.2 | 2337.0 | 20.00 |
| 0 | Inner surface | 1402.2 | 2192.1 | 18.97 |
| 1 | Limit between layers 1/2 | 1373.5 | 2176.1 | 18.85 |
| 2 | Limit between layers 2/3 | 521.9 | 1585.6 | 13.88 |
| 3 | Limit between layers 3/4 | 251.5 | 251.5 | -10.34 |
| 20 | Outer surface | 205.5 | 249.3 | -10.44 |
| | Outdoor air | 205.5 | 241.8 | -10.78 |

The outside air temperature during condensation is -17.78°C. Table 5 shows the indoor and outdoor climatic parameters during evaporation. It is observed that the outside air temperature during evaporation is 9°C.

**Table 5.** The amount of condensed water in the cold season.

| | | | | |
|---|---|---|---|---|
| **Indoor climatic parameters** | | | | |
| Indoor air temperature | | | 20 | °C |
| Relative humidity of indoor air | | | 60 | % |
| Heat transfer coefficient at the inner surface | | | 8 | W/m²°C |
| **External climatic parameters** | | | | |
| **External eir temperature during condensation** | | | **-17.78** | **°C** |
| Relative humidity of outdoor air | | | 85 | % |
| Coefficient of thermal transfer of the outer surface | | | 24 | W/m²°C |

| Crt. No. | Layers surfaces | Partial pressure (Pa) | Saturation pressure (Pa) | Temperature (°C) |
|---|---|---|---|---|
| Indoor air | | 1402.2 | 2337.0 | 20.00 |
| 0 | Inner surface | 1402.2 | 2160.3 | 18.74 |
| 1 | Limit between layers 1 / 2 | 1371.2 | 2140.9 | 18.59 |
| 2 | Limit between layers 2 / 3 | 450.0 | 1447.9 | 12.49 |
| 3 | Limit between layers 3 / 4 | 157.6 | 133.5 | -17.24 |
| 4 | Limit between layers 4 / 5 | | | |
| 20 | Outer surface | 107.9 | 132.1 | -17.36 |
| | Outdoor air | 107.9 | 126.9 | -17.78 |

| **Parameters for calculating the amount of condensed water** | | |
|---|---|---|
| Saturation pressure at the first tangent point | 133.5 | Pa |
| Saturation pressure at the second tangent point | 133.5 | Pa |
| Vapor permeability resistance $R'_v / 10^8$ | 114.9 | m/s |
| Vapor permeability resistance $R''_v / 10^8$ | 4.6 | m/s |
| Duration of the condensation process | 772.0 | hours |
| **The amount of water accumulated by condensation** | **0.1515** | **Kg/m²** |

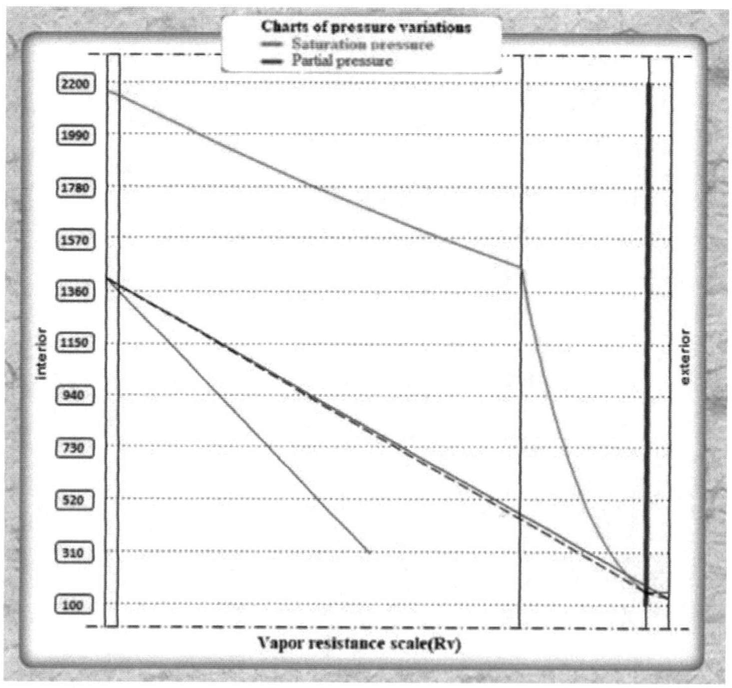


**Figure 4.** Diagram of pressure variation graphs.

It is observed that there is no risk of water accumulation – figure 4 and table 6 by checking successive wetting.

**Table 6.** Indoor and outdoor climatic parameters during evaporation.

| | | | | |
|---|---|---|---|---|
| **Internal climatic parameters** | | | | |
| Indoor air temperature | | | 20 | $^0C$ |
| Relative humidity of indoor air | | | 60 | % |
| Heat transfer coefficient at the inner surface | | | 8 | $W/m^{20}C$ |
| **External climatic parameters** | | | | |
| **External air temperature during condensation** | | | **9.00** | **°C** |
| Relative humidity of outdoor air | | | 70 | % |
| Coefficient of thermal transfer of the outer surface | | | 24 | $W/m^{20}C$ |
| **Crt. No.** | **Layers surfaces** | **Partial pressure** (Pa) | **Saturation pressure** (Pa) | **Temperature** (°C) |
| | Indoor air | 1402.2 | 2337.0 | 20.00 |
| 0 | Inner surface | 1402.2 | 2284.3 | 19.63 |
| 1 | Limit between layers 1 / 2 | 1387.8 | 2278.3 | 19.59 |
| 2 | Limit between layers 2 / 3 | 961.5 | 2038.8 | 17.81 |
| 3 | Limit between layers 3 / 4 | 826.2 | 1159.6 | 9.16 |
| 4 | Limit between layers 4 / 5 | | | |
| 20 | Outer surface | 803.2 | 1157.0 | 9.12 |
| | Outdoor air | 803.2 | 1147.5 | 9.00 |
| **Parameters for calculating the amount of water evaporated** | | | | |
| Saturation pressure at the beginning of the condensation zone | | | 1159.6 | Pa |
| Saturation pressure at the end of the condensation zone | | | 1159.6 | Pa |
| Vapor permeability resistance $R'_v / 10^8$ | | | 114.9 | m/s |
| Vapor permeability resistance $R''_v / 10^8$ | | | 4.6 | m/s |
| Duration of the evaporation process | | | 7988.0 | ore |
| **The amount of water removed by evaporation** | | | **21.7177** | **$Kg/m^2$** |

**Table 7.** Calculation parameters for checking excessive moisture.

| **Calculation parameters for checking excessive moisture** | | |
|---|---|---|
| The amount of water vapor that can condense in the element during the cold period of the year | 0.1515 | $Kg/m^2$ |
| Density of material moistened by condensation | 20.00 | $Kg/m^3$ |
| The thickness of the layer of material in which the accumulation of water occurs | 0.10 | m |
| **Increase in mass humidity at the end of the condensation period** | **3.60** | **%** |
| **Maximum permissible increase in mass humidity** | **15.00** | **%** |

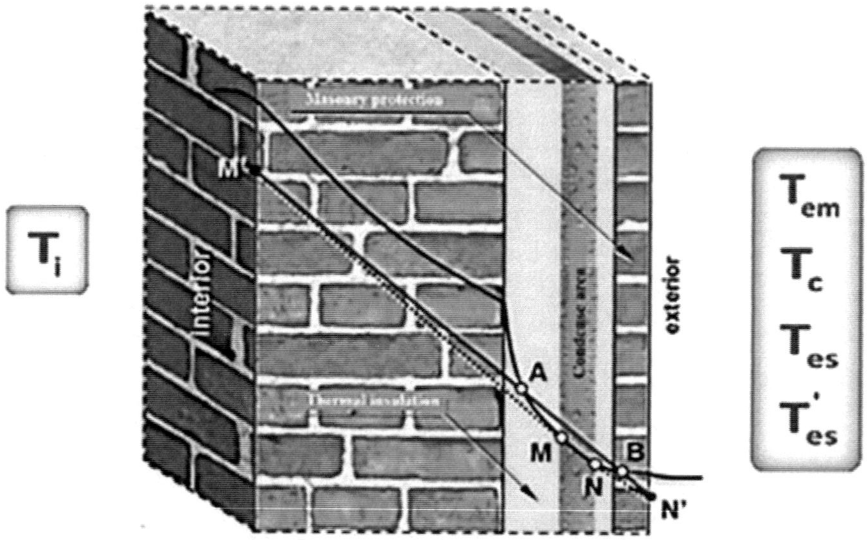


**Figure 5.** Condensate risk assessment for cork thermal insulation

Figure 5 makes a synthesis of the mass transfer analysis for condensate risk assessment for cork thermal insulation. The average outside air temperature during condensation is -13.38°C. The amount of condensed water accumulated in the cold season is 0.0720kg/m$^2$. The amount of water eliminated by evaporation during the hot season is 20.2460kg/m$^2$. Increased humidity at the end of the condensation period is 3.60% (maximum admissible being 15%).

Regarding the energy consumption to ensure comfort, it was done in figure 6 a comparison between a classically insulated building with 10 cm VMB and one in which the thermal insulation is made of cork of the same thickness. The model building from climate zone IV described above was used.

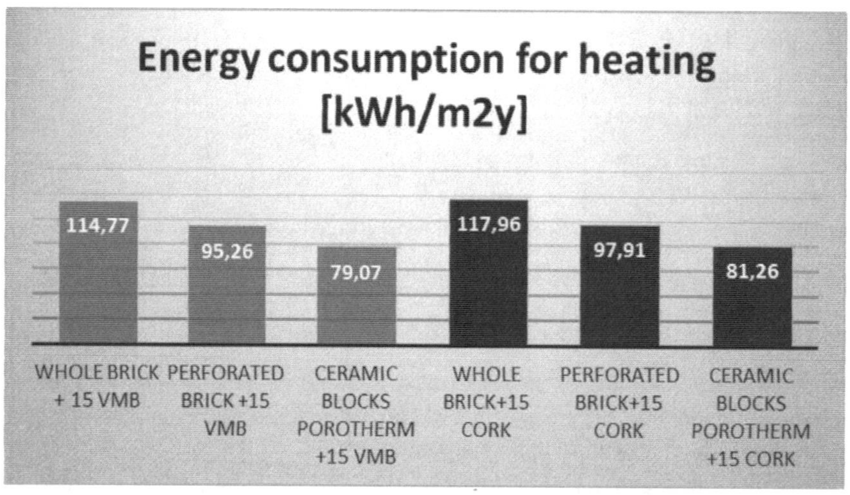


**Figure 6.** Comparative values for heating energy consumption that ensure interior comfort using classic insulation and cork.

Although the differences in energy consumed for heating are quite small because the values of thermal conductivity are very close, the cork has a high specific thermal capacity and a high thermal inertia compared to the classical insulation. It behaves well in a wide range of temperatures - keeping houses pleasantly warm in winter and cool in summer.

Cork provides good sound insulation in addition to its excellent performance as a thermal insulator. The structure and flexibility allow it to absorb up to 70% of the tones that are in the frequency range 400-4000Hz. These soundproofing and anti-vibration properties help to transform cork into an ideal building material. It works especially well for urban areas where there is a strong desire to turn off the noise of the crowded outside world.

Also, regarding the comparison between the costs necessary for thermal insulation in the classic version and the one using cork as thermal insulation, it can be seen from figure 7, the cork has a higher cost/sqm than that of the classic insulator.

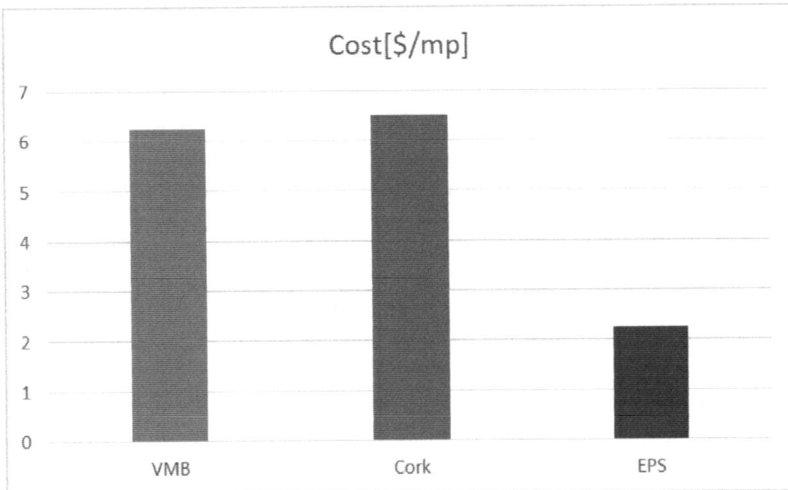


**Figure 7.** Comparative values for the unit cost of classical insulation and cork.

Although the initial investment is higher, the cushioning is done quickly enough, and the interior comfort is increased since its use prevents common problems, such as mold and rot.

## 4. Conclusions

Ensuring the fundamental requirement applicable to constructions related to hygiene, health and environment as well as ensuring the fundamental requirement regarding energy saving and thermal insulation brings the evaluation of hygroscopic characteristics as an element of major importance for research and innovation in insulating materials for buildings.

The study presented is useful both for designers for buildings in order to efficiently choose the composition of the building's envelope closure elements, thermal calculation and calculation of water vapor diffusion and manufacturers of construction materials that must indicate in the Product Declarations of Performance, the characteristics of thermal conductivity / thermal resistance and water vapor permeability. The realization of such closing elements that have a favorable behavior to the diffusion of water vapor satisfies the requirement of sustainability of constructions which represents an essential component of the fundamental requirement regarding the sustainable use of natural resources.

Cork is an efficient high-performance thermal insulator compared to the thermal insulation used on a large scale. As the study shows, cork helps prevent condensation. This type of thermal insulation has diffusion capacity, ie it does not influence the release of moisture. Considering the fact that it is a very good sound insulator, this product can be recommended as an ecological product (green) excellent for thermo-hygro-sound insulation of residential homes.

IOP Conf. Series: Earth and Environmental Science **960** (2022) 012005 doi:10.1088/1755-1315/960/1/012005

All the conclusions of this research can be used in designing passive houses and green technologies for passive houses in order to achieve all the performances requested to these type of future buildings.

## References

[1] Tabrea A 1998 *Alcătuiri îmbunătăţite de pereţi şi acoperişuri din punct de vedere higrotermic,* Teză de doctorat, Universitatea Tehnică de Construcţii Bucureşti, *Improved composition of walls and roofs from a hygrothermal point of view*, Doctoral thesis, Technical University of Constructions Bucharest

[2] fenix.tecnico.ulisboa.pt < downloadFile The nZEB buildings: Cork, as case study Técnico Lisboa

[3] http://www.isocor.pt/aplicacoes

[4] https://www.amorimcorkinsulation.com/en/advantages

[5] Zhang B Q and Liu X F 2003 Effects of Fractal Trajectory on Gas Diffusion in Porous Media, *AIChE Journal* **49**(12) 3037-47

[6] Calbureanu M X, Albota E, Tutunea D, Dumitru S, Malciu R and Dima A 2010 Contributions above the dew-point problem in civil building EPS insulated walls modeling with finite element the convective heat transfer, *International Journal of Mechanics* **3(4)** 53-62

[7] Calbureanu M, Talu M, Travieso-González C M, Talu S, Lungu M and Malciu R 2010 The finite element analysis of water vapor diffusion in a brick with vertical holes, *International Conference on Mathematical Models for Engineering Science (MMES '10),* Puerto De La Cruz Tenerife 57-62

[8] Incropera F P and Witt D P 1996 *Fundamentals of Heat and Mass Transfer* (New York: John Willey &Sons).

[9] C 107/0 2002 Normativ pentru proiectarea şi execuţia lucrărilor de izolaţii termice la clădiri

[10] C 107 2005 Normativ privind calculul termotehnic al elementelor de construcţie ale clădirilor

[11] SR EN 12086:2013 Produse termoizolante destinate utilizării la clădiri. Determinarea proprietăţilor de transmisie a vaporilor de apă

[12] https://ecoesmas.com/el-corcho-los-paneles-de-corcho-y-sus-ventajas/

# A Numerical Approach of the Influence of Water Temperature in the Extinction of Fire Developed in Closed Spaces

**P Cozma[1, 3], D Pavel[2], C A Safta[3] and A Chisacof [3]**

[1] The General Inspectorate for Emergency Situations, Bucharest, Romania
[2] Police Academy, Firefighters Faculty, Bucharest, Romania
[3] University Politehnica of Bucharest, Power Engineering Faculty, Romania

E-mail: safta.carmenanca@gmail.com;
ORCID: https://orcid.org/0000-0002-4651-0570

**Abstract.** The aim of this work is to make a visual investigation of the fire behavior during the extinguish process with water mist at different temperatures, different diameters of the nozzle and different angles of diffusion jet, too. It was used a numerical simulation approach to analyze 24 scenarios regarding the influence of mist water in fire extinguish process. The main test was focused on a given enclosure with dimensions of 1.0 × 1.0 × 3.0 m and the fire source placed inside, in the middle. The enclosure is considered opened on the upper part. From the numerical simulation values, we concluded the best scenario in fire extinguish process using water mist.

## 1. Introduction

Life safety and personal protection, incident stabilization and property conservation are three important items which must be accomplished in any extinguish action [1]. The property conservation depends on the time in which the fire can be suppressed, and the amount of agent extinguished used. When the agent extinguish is water, the amount of water is very important because water is one of the common extinguish agent used to remove the heat of the fire. In these circumstances, in the last decades the water mist extinguisher systems were developed, by using de-ionized water discharged in a fine spray of droplets onto the burning material [1].

Based on a previous experimental approach [2], in this paper is developed a numerical simulation study regarding the influence of water temperature in the process of fire suppression. It was used the Fire Dynamics Simulator (FDS) having PyroSim as graphical user interface.

Computer simulations of fires is a well-known topic in the literature survey [3-14] in which PyroSim is used in simulation of the fire and smoke evolution in fire safety actions in tunnels and underground tunnels [3-5], performance room [6], and inside/outside of different structures [7-13]. PyroSim and Pathfinders are powerful instruments in numerical simulation technique offering the possibility of planning a fire rescue action [10-12].

FDS models the low-speed thermal flow and conduction of physical and chemical processes associated with fire and smoke spread. In PyroSim can be implemented in the 3D structure different types of devices accustomed in fire security analyzes as heat detectors, thermocouples, pressure sensors, ventilation system, sprinklers, and nozzles. So, a variety of "fire scenes" can be created for numerical simulations [3, 4, 7].

In the literature are technical reports and papers that provide an important material for the validation of the FDS fire simulation program and eligible material parameter values [8, 14].

The aim of this paper is a numerical analyze of the influence of water used as extinguish agent in water mist extinguishment method. There are considered 24 scenarios that depend on water temperature, the angle of diffusion jet in the nozzle and the nozzle diameter. In each scenario, the nozzle is placed above the fireplace at the same distance, and in the same condition of enclosure. From this study is found the type of nozzle and the temperature of water in which the fire is suppressed with a better efficiency.

The paper describes the methodology used and the numerical simulation results.

## 2. Methodology

Based on the experiments presented in paper [2] we are considering the fire feed by methane gas and the water mist as agent extinguish at the pressure of the domestic water network, at different temperatures between 20 °C to 40 °C.

The geometry of the numerical simulation model is considered a 3D rectangular encloser with a square floor 1 × 1 m and a height of 3 m. The material of walls was considered an inert material.

The upper part of the encloser is open to have the necessary amount of oxygen in the burning process. In the middle of the floor is placed the fire burner with a methane gas having the heat release rate per unit area HRRPU of 50 kW/m$^2$, and the burning surface area of 0.1 × 0.1 m.

Above the fire burner it was placed a nozzle on 0.15 m on the same vertical axis with the burner which was keeping constant during all the scenarios considered. In figures 1 and 2 are presented the sketch of the experiment and the 3D geometrical representation of the encloser.

In the numerical experiments were used nozzles having two types of jet diffusion angles, $\alpha =$ (45°; 60°), and four different diameters $d = (1.25; 1.5; 2.00; 2.25)$ mm. In figure 3 is drawing the nozzle with the geometrical elements considered to vary.

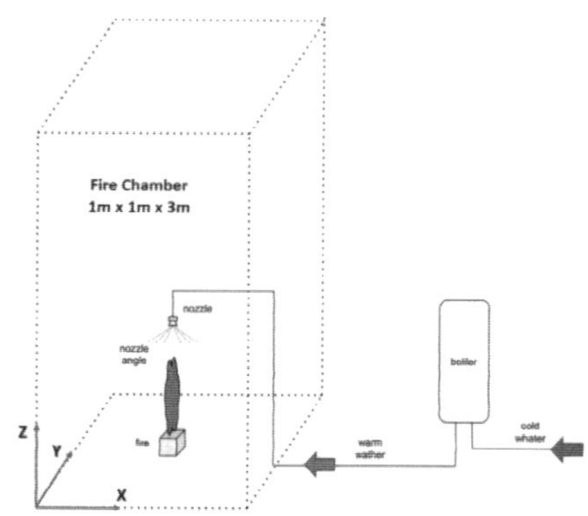


Figure 1. The experiments description.

Figure 2. The numerical grid of the fire encloser.

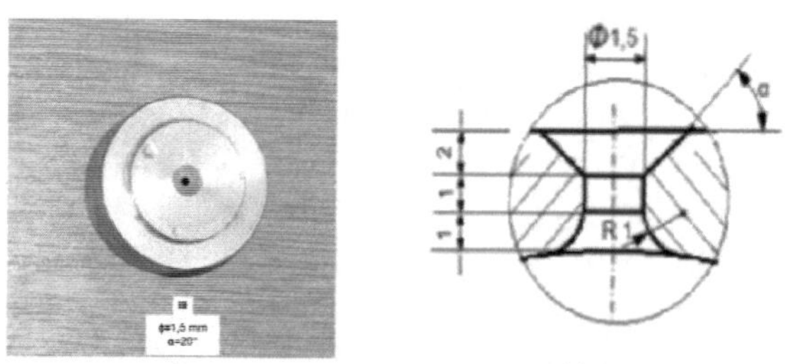


a) b)

**Figure 3.** The dispersion head nozzle (a), nozzle detail (b) with diameter $d = 1.5$ mm and angle $\alpha = 45°$.

**Table 1.** Input data for numerical simulation [15].

| Category | Parameters | Value |
|---|---|---|
| | Simulation volume | Fire chamber, designed in the program |
| Computational domain | Dimension of the chamber | 1 m × 1 m × 3 m (in order on 0X, 0Y and 0Z axis) |
| Numerical data | Cells per mesh, uniform division | 32 × 32 × 48 cells |
| | Dimension of one cell | 0.06 m × 0.06 m × 0.06 m |
| | The total number of cells | 49152 |
| | Boundary conditions for the simulation volume | The boundary of the domain was considered closed, except for the plane Z=3 m with bounds min X = -1 m, max X = 1 m and min Y = -1 m, max Y = 1 m, which was considered open. The whole structure of the chamber was considered inert. |
| Other data | The fire burner | Imported from the FDS library, HRR (heat release rate) of 50 kW/m$^2$, square shape fire source, dimensions 0.1 m × 0.1 m |
| | Environment values | 20 °C, relative humidity 40 %, atmospheric pressure 101325.00 Pa, ground level at 500 m about sea level |
| | Simulation time | 20 seconds |
| | Nozzle | Imported from FDS, flow 5 L/min, *k* factor=1, droplet median diameter 8000 μm |

Heated water at temperatures of (20; 30; 40)°C was used as extinguish agent, for each type of nozzle.

The nozzle is activated in 4 seconds after the combustion is started. The water is discharged into the nozzle at a flow rate of 5 L/min. The water mist was defined as having 8000 drops/second, and the average diameter of the drop was considered the combination of 300 μm and volume given by the Rosin-Rammler function (gamma distribution 2.4; minimum diameter 20 μm by default; maximum diameter, infinite by default) [15]. Minimum diameters are assumed to evaporate in a single time step, eliminating numerical instabilities that can occur when droplets get very, very small. The water density was considered 1000 kg/m$^3$ and the specific heat at constant pressure equal to 4184 J/kgK.

According to the simulation data [14, 15], the temperature fields developed the gradient on the vertical up to 50 °C at around 1 m height. For this reason, it was placed a thermocouple at 1 m vertical distance from the burner. It was monitoring the temperature developed during the burning process.

The input data for numerical parameters are presented in table 1.

The mesh size was chosen considering the condition given by the User's Guide [15]:

$$\frac{D^*}{\partial x} \in (4 \div 16) \tag{1}$$

with $\partial x$ is the nominal size of the mesh cell.
The characteristic diameter of fire $D^*$ was calculated with:

$$D^* = \left(\frac{Q}{\rho_\infty c_p T_\infty \sqrt{g}}\right)^{2/5} = \left(\frac{50}{1.177 \cdot 1.0049 \cdot 300\sqrt{9.81}}\right)^{2/5} = 0.289 \text{ m}, \tag{2}$$

where: $\rho_\infty$ is air density, $c_p$ – air specific heat, $T_\infty$ – temperature of the environment, $Q$ – heat release rate of the burner fuel. For the limits given in relation (1) the nominal size of mesh cell can vary in $(0.018 \div 0.072)$ m.

PyroSim is using the numerical model based on mass conservation equation, momentum conservation and energy conservation equations.

The numerical method used is large eddy simulation, LES.

## 3. Numerical simulation results

The scenarios analyzed in the paper were named as follows SC **/##/@@ with ** the angle of jet diffusion $\alpha$, ## the diameter of the nozzle $d$, @@ the water temperature, e.g., SC 45/1.25/30 is the scenario with $\alpha = 45°$, $d = 1.25$ mm and 30 °C water temperature.

For the data given in table 1 and previously described the temperature during the burning process, the heat release rate, and the heat flow were monitored. Also, the temperature distribution is plotted at the 6th second, 12th second and 20th second of burning evolution. In figure 4 the temperature evolution of the process in case of encloser without any extinguish system is presented, and in figure 5 is the evolution of temperature, heat release rate and heat flow are illustrated for this this case, which can be considered case zero SC 00/00/00.

PyroSim automatically calculates the amount of smoke generated but in this case the amount of smoke is low having methane as flame burner.

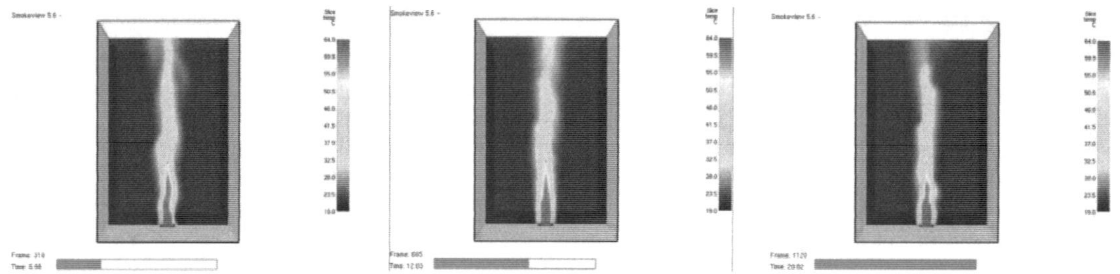

a) b) c)

**Figure 4.** Evolution of fire in the encloser without any extinguish system: a) 6 s; b) 12 s; c) 20 s

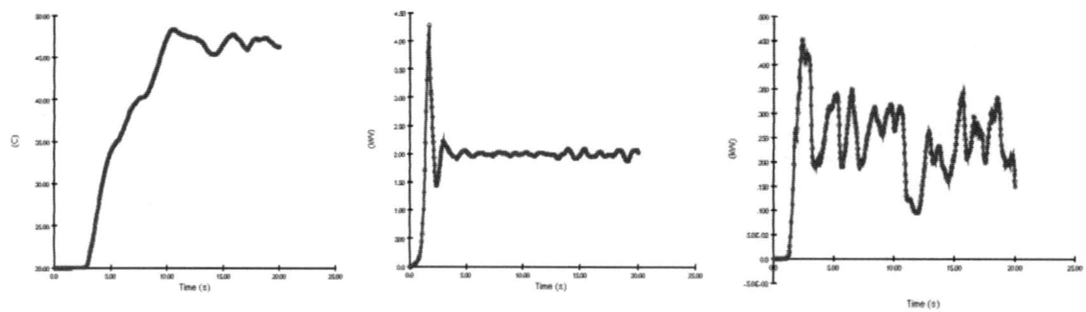

a) b) c)

**Figure 5.** Evolution in time of the temperature (a), heat release rate (b) and heat flow (c) in case of the fire without any extinguish system.

The numerical results were analyzed considering the temperature evolution monitoring with the thermocouple mounted at 1 m above the burner on the same vertical axis with the nozzle. In figure 6 are given the parameter considered in the analysis: $T_{max}$ the maximum temperature, $t_{Tmax}$ the time in which the maximum temperature has been reached, $T_{20}$ the temperature in the 20th second of simulation, and $t_d$ is time delay. It is observed that the time delay has had the same value for all scenario, $t_d = 2.67$ s.

In figures 7 ÷9 the parameters mentioned in figure 6 are presented for all analyzed scenarios.

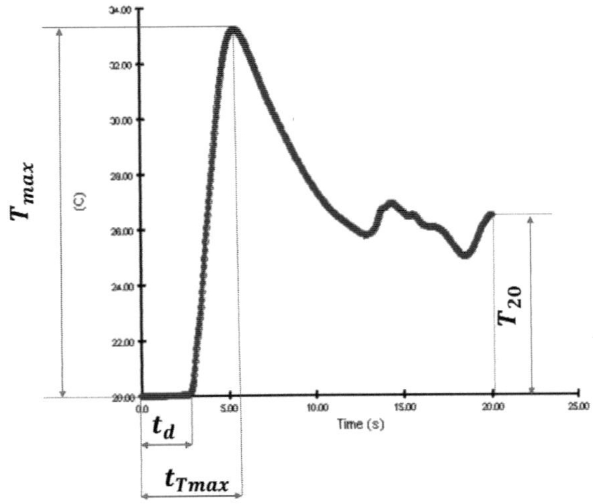


**Figure 6.** Characteristic parameters analysed in the temperature evolution.

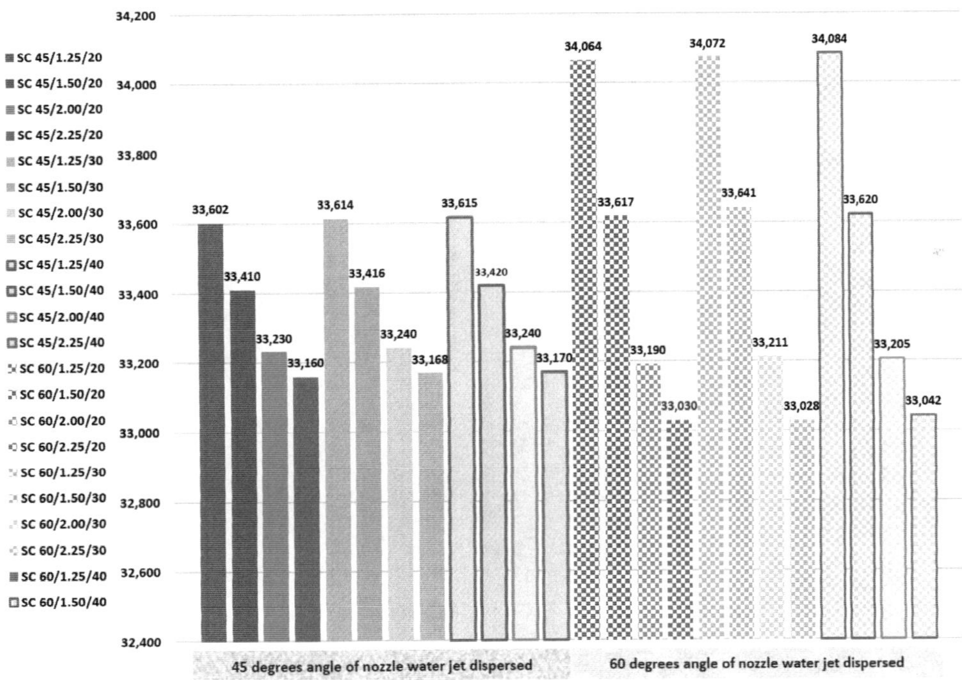


**Figure 7.** Maximum temperature, $T_{max}$, evolution on all scenarios analysed.

It is observed that maximum value is $T_{max}$ = 34.084 °C for SC 60/1.25/20 and the minimum value is $T_{max}$ = 33.028 °C for SC 60/2.25/30. Also, the lowest values of $T_{max}$ are for $d$ = 2.25 mm no mater $\alpha$ angle, and water temperature. Regarding the water jet diffusion angle, it is observed that $T_{max}$ has lowest values on $d$ = 1.25 mm and $d$ = 1.50 mm for $\alpha = 45°$ by comparison with the value obtained on the same diameters but with $\alpha = 60°$. Very interesting, $T_{max}$ values are lowest for $d$ = 2.00 mm and $d$ = 2.250 mm if the angle $\alpha = 60°$ by comparison with $\alpha = 45°$. The graphical representation of $T_{max}$ shows that the values are decreasing if the temperature of water is increasing whatever the diameter and water diffusion jet angle is.

IOP Conf. Series: Earth and Environmental Science **960** (2022) 012006 doi:10.1088/1755-1315/960/1/012006

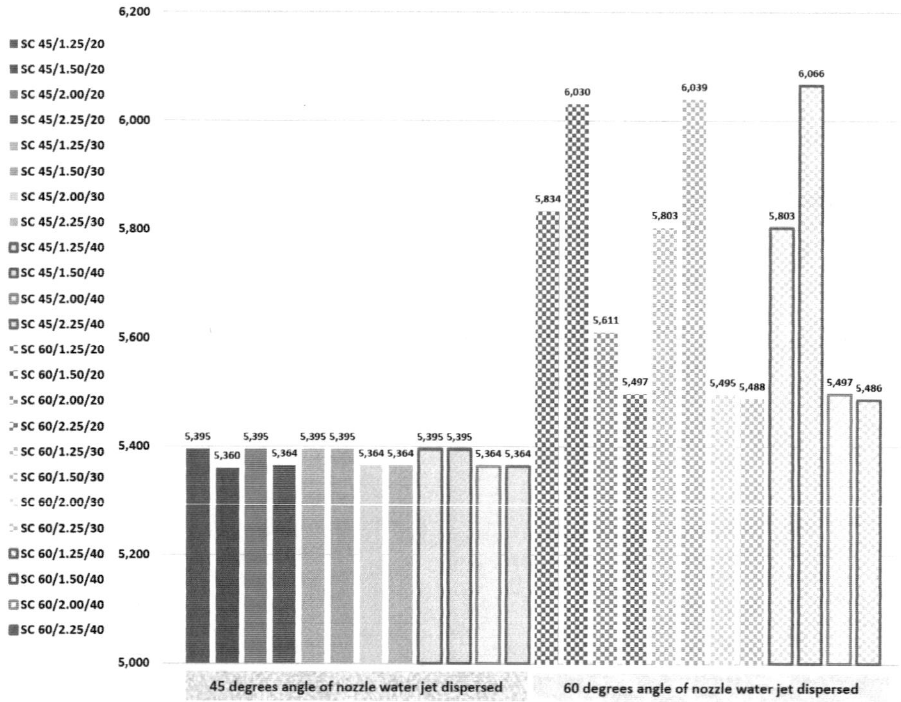


**Figure 8.** The time $t_{Tmax}$ in which the temperature is maximum $T_{max}$.

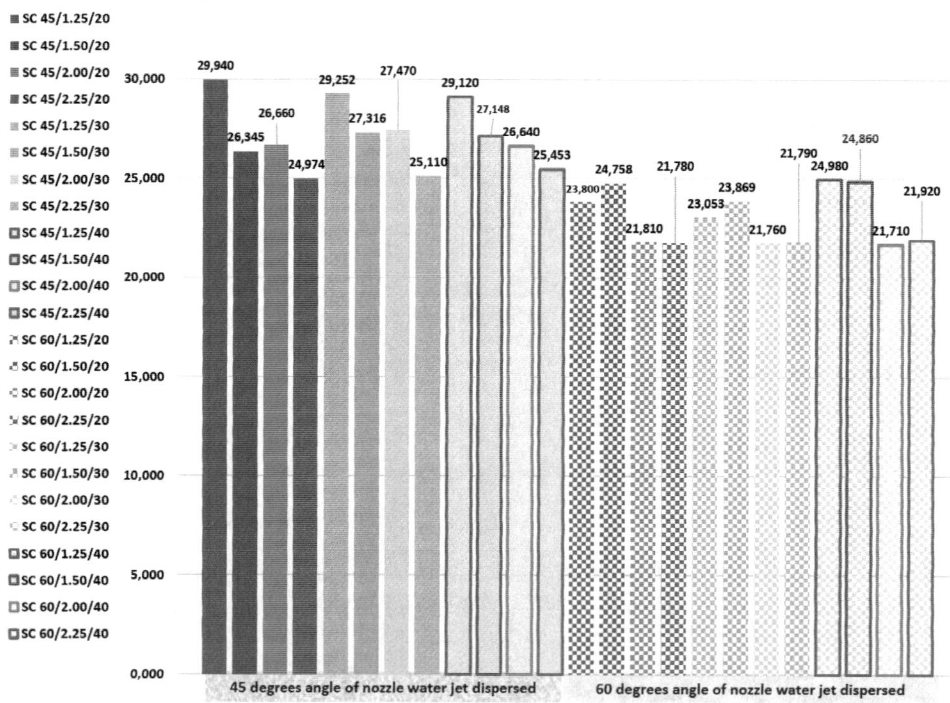


**Figure 9.** Temperatures in the 20th second of the simulation, $T_{20}$ , on all scenarios.

Analyzing the results plotted in figure 8 it can be concluded that the lowest $t_{Tmax}$ values are for $\alpha = 45°$ and not for $\alpha = 60°$ no matter the nozzle diameter and the water temperature. According as the water temperature is increasing the times $t_{Tmax}$ can be lower or equal between them when the diameters are 2.00 mm and 2.25 mm in both cases of $\alpha = 45°$ and $\alpha = 60°$. Also, the lowest time is 5.360 s in the case of scenario SC 45/1.50/20 and the highest value is 6.066 s for SC 60/1.50/40.

In figure 9 the lowest value of $T_{20}$ is 21.71 °C for scenario SC 60/2.00/40 and the highest is 29.94 °C for SC 45/1.25/20. It is observed that with the increasing of water temperature, the values of $T_{20}$ are decreasing no matter the nozzle diameter and the angle of water diffusion jet. On the other hand, for the same diameters but different $\alpha$ angles $T_{20}$ is lowest at $\alpha = 60°$ than for $\alpha = 45°$. Also, at the same $\alpha$ angle but different diameters the lowest $T_{20}$ is for $d = 2.00$ mm and $d = 2.25$ mm.

From the analyzed data it seems that SC 60/2.00/40 is the optimum configuration of nozzle and water temperature if we are considering as the efficiency of the system, the temperatures ratio:

$$\eta_T = \frac{T_{max} - T_{20}}{T_{max}} \times 100. \tag{3}$$

Was calculated $\eta_T$ for all scenarios and the highest value is 34.6% for SC 60/2.00/40 and the lowest is 10.9% for SC 45/1.25/20.

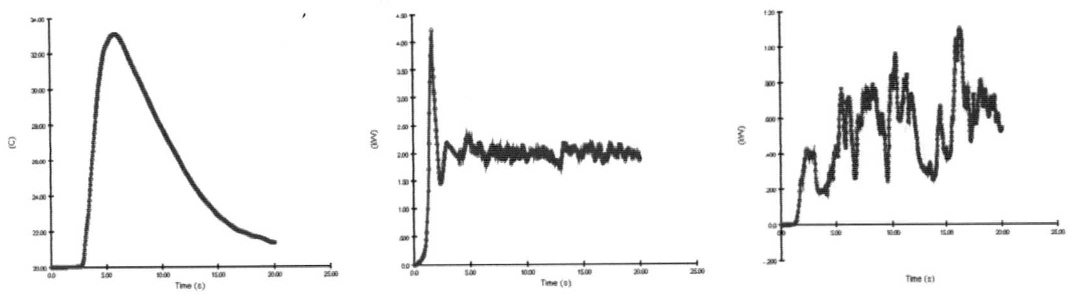

a) b) c)

**Figure 10.** Evolution in time of the temperature (a), heat release rate (b) and heat flow (c) in case of the fire with the extinguish system SC 60/2.00/40.

**Table 2.** Parameters in fire evolution for the optimum configuration of nozzle and water temperature compared with fire.

| Parameter | Unit | SC 60/2.00/40 | Fire |
|---|---|---|---|
| $T_{max}$ | °C | 33.205 | 48.397 |
| $T_{20}$ | °C | 21.71 | 46.335 |
| $HRR_{max}$ | kW | 4.224 | 4.434 |
| $HRR_{20}$ | kW | 2.104 | 2.03 |
| $HF_{max}$ | kW | 0.42709 | 0.4531 |
| $HF_{20}$ | kW | 0.2048 | 0.1501 |

In table 2 are listed all the parameters involved in fire process for scenario SC 60/2.00/40 to be compared with parameters of scenario SC 00/00/00 fire without any extinguish system. It was monitored the temperature ($T_{max}$, $T_{20}$) and the extreme values obtained for the heat release rate $HRR$ and heat flow $HF$. The values denoted $HRR_{20}$, $HF_{20}$ are obtained in the last second of simulation time.

## 4. Conclusions

Using PyroSim as numerical simulation environment the paper presents the analyses of the influence of water mist temperature used as extinguish agent in the suppress process of a methane burner.

It was described the methodology used and the numerical simulation results for 24 scenarios of nozzle water mist were discussed. The numerical simulation results underline that increasing the temperature of the water mist in the extinguish process can improve the fire suppression. Also, it was found the optimum configuration of nozzle and water mist temperature.

The future work will be conducted to validate the numerical results on a full-scale configuration.

**References**

[1] NFPA Fire Protection Handbook 20th Edition.
[2] Pavel D, Cozma P, Dimitriu S, Chisacof A and Safta C A 2021, Mixture Between Fire and Mist Jets Characteristics for Flame Extinguish, IOP Conference Series Earth and Environmental Science 664(1):012048, DOI: 10.1088/1755-1315/664/1/012048
[3] Adjiski V, Mirakovski D, Despodov Z and Mijalkovski S 2015, Simulation and optimization of evacuation routes in case of fire in underground mines, J. of Sustainable mining 14, 133-143, http://dx.doi.org/10.1016/j.jsm.2015.10.001
[4] Codescu S, Popescu D and Panaitescu V 2015, Computational simulation of road tunnel fire protection by using sprinklers, U.P.B. Sci. Bull., Series D, Vol. 77, Iss. 2, 2015, ISSN 1454-2358, pp. 271-278.
[5] Xiao Ju Li et al 2018 IOP Conf. Ser.: Earth Environ. Sci. 153 062032
[6] Valasek L 2013, The use of PyroSim graphical user interface for FDS simulation of a cinema fire, Int. J. of Mathematics and computer simulation, Issue 3, Volume 7, 2013, pp 258-266.
[7] Li K Y, Spearpoint M and Fleischmann C 2010, Numerical Simulation of Smoke Downdrag due to a Sprinkler Spray using FDS, Proc. 8th Asia-Oceania Symposium on Fire Science and Technology (AOSFST), Melbourne, Australia, 7 – 9 Dec 2010, https://ir.canterbury.ac.nz.
[8] Nedryshkin O, Gravit M and Grabovyy K 2018, Modeling fires in structures with an atrium in the FDS field model, MATEC Web of Conferences 193, 03023, ESCI, 2018, https://doi.org/10.1051/matecconf/201819303023
[9] Simion A et al 2018 IOP Conf. Ser.: Mater. Sci. Eng. 374 012082
[10] Long X, Zhang X and Lou B 2017, Numerical simulation of dormitory building fire and personnel escape based on Pyrosim and Pathfinder, Journal of the Chinese Institute of Engineers, DOI: 10.1080/02533839.2017.1300072
[11] AIP Conference Proceedings 1980, 050026 (2018); doi: 10.1063/1.5044362
[12] Hu G 2020 IOP Conf. Ser.: Earth Environ. Sci. 455 012059
[13] Zhang Q et al 2020 IOP Conf. Ser.: Earth Environ. Sci. 565 012090
[14] Hietaniemi J, Hostikka S and Vaari J 2004, FDS simulation of fire spread- comparison of model results with experimental data, VTT Working Papers 1459-7683, VTT Building and Transport, Finland, 2004.
[15] McGrattan K, Klein B, Hostikka S and Floyd J 2007, Fire *Dynamics Simulator (Version 5) User's Guide,* , NIST Special Publication 1019-5, U.S. Government Printing Office Washington.

IOP Conf. Series: Earth and Environmental Science **960** (2022) 012007 doi:10.1088/1755-1315/960/1/012007

# Composite materials based on natural fibres applied for structural reinforcement

**G Bou Abdallah[1,2], I Ivanova[1*], J Assih[2], C Diagana[2] and D Dontchev[1]**
[1]Department of Applied mechanic, University of Chemical Technology and Metallurgy, Kliment Ohridski Boulevard, 8 Sofia, 1756
[2] Civil Engineering Department Sustainable Construction, ITHEMM EA 7548, University of Reims Champagne-Ardenne

E-mail: i.ivanova@uctm.edu
ORCID: https://orcid.org/0000-0001-9332-5848

**Abstract**. Environmental problems and environmental protection triggered a rapid development of natural fibres as sustainable materials for the reinforcement of reinforced concrete structures. Synthetic fibre polymer composite materials have been widely accepted by the construction industries as an effective external reinforcement material to rehabilitate deficiencies in existing structures. These materials have exceptional performance such as high strength to weight ratio, corrosion resistance and lightness. However, the disadvantages include high costs during manufacturing and end-of-life services, less environmentally friendly and causing adverse effects on human health. This article presents an experimental program on the use of natural fibres as reinforcement in composite materials for structural strengthening. Different types of natural fibre fabrics (hemp, flax, mixed hemp and cotton) in terms of their mechanical properties were studied. The fibre and fibre fabric sheets were tested in tension test and compared with carbon and glass fibre fabric sheet as reference. So, this study carries out the effect of natural hemp and flax fibre fabric thickness on ultimate loads of specimens. In addition, the ultimate load and stiffness of strengthened beams were investigated. In fact, the results show that the reinforcement technique allows to increase the load-bearing of strengthened structure by 8% to 35% in bending tests.

## 1. Introduction

In recent years, many works have been carried out to reduce pollution and its impact on the environment. The use of composite materials based on natural fibre remains is a major challenge for scientists. In fact, the use of sustainable natural materials would considerably reduce harmful pollutant emissions. The natural plant fibres from naturally growing plants as a reinforcement in composite materials reduces cost and increases biodegradability of the products hence to decrease environmental pollution.

These materials have many advantages such as renewable sources, biodegradability, which leads to low environmental impact, hence reducing environmental pollution, reduces costs, and their non-corrosive properties. The fields of application of these materials are numerous: in civil engineering, aircraft construction, automotive, marine structures, sports equipment, textiles. These materials have exceptional performance such as high strength to weight ratio, corrosion resistance and lightness. However, the disadvantages include high costs during manufacturing and end-of-life services, less environmentally friendly and causing adverse effects on human health. And they have interesting mechanical properties like the light weight and the ability to manufacture complex shapes. The effectiveness of the technique of reinforcement by the bonding of composite materials is one of the most significant advantages justifying more their applications in industry.

The synthetic fibre polymer composite materials are widely used in the construction industries as an effective external reinforcement material to rehabilitate buildings or constructions [1-4]. *But* in 2018 after the adoption of Farm Bill, the hemp was becoming a federally legal crop in the United States for the first time in nearly a half century.

The main disadvantage of using natural fibres as reinforcement of polymeric matrices is the physicochemical incompatibility between the fibres and the matrix. Because of the hydrophilic nature of fibres and the hydrophobic behaviour of resins [5]. To improve the properties of composites it is necessary to treat both materials.

Ren et al. [6] investigated the use of natural hemp fibre as reinforcement for the possibility of producing fibre reinforced Sheet Molding Compounds (SMC) as an alternative to glass fibres in the industrial applications ranging from building construction, automotive, to aerospace, where a good fire performance and mechanical properties are essential. Their results show that natural fibre SMC achieved equivalent level of fire and mechanical properties that glass fibre SMC.

The aim of this experimental study is to study the mechanical properties of natural fibres as reinforcement composite material in strengthened structure. The objective is to replace the polymer composites materials with natural materials as they are environmentally friendly and they are not dangerous to human health. The novelty in this research is the study of natural fibres as a reinforcing element in concrete structures.

In the development of this article, the materials will be presented and the investigation through the influence of the thickness of the composite plate and the effect of the reinforcement on ultimate load of strengthened beams.

## 2. Experimental program

### *2.1. Materials*

This leads to an educational campaign on students to knowledge of and attitude toward industrial hemp. They find that many of them do not know the difference between hemp and marijuana [7]. The hemp fibres are used for removal of heavy metals from aquatic ecosystems, including lead ions. In contrast, hemp is a commonly available unconventional material that can be efficiently used for heavy metal removal from wastewaters. The performances of hemp as sorbent are based on its remarkable fundamental features: high porosity, hydrophilic character, fast sorption, tolerance to biological structures, easiness in functionalization, possibility of being used as fibres and filters [8], [9]. The mechanical and physical properties of natural fibres are shown in Table 1 below.

**Table 1.** Average tensile mechanical proprieties of different natural fibres compared to reinforcing fibres usually used in the composite materials industry [10].

| Fibres | E (GPa) | A (%) | $\sigma_u$ (MPa) | Density |
|---|---|---|---|---|
| | | **Industry fibres** | | |
| **Glass E** | | | | |
| Virgin filament | 72 à73 | 4,6 à 4,8 | 3 200 à 3 400 | 2,54 |
| Industrial filament (basic yarn with a large number of fibres | 72 à73 | 1,7 à 2,1 | 1 200 à 1 550 | 2,54 |
| **Carbon** | | | | |
| Toray T300 | 230 | 1,5 | 3 530 | 1,7 à 1,9 |
| Thorneel P-120 S | 825 | 0,3 | 2 350 | 1.87-2 |
| **Aramid** | | | | |

| | | | | |
|---|---|---|---|---|
| Kevlar 49 | 124 | 2,9 | 3620 | 1,44 |
| | | **Vegetable fibres** | | |
| **Flax** | | | | |
| general data | 12 à85 | 1 à 4 | 600 à 2 000 | 1,54 |
| **Hemp** | | | | |
| general data | 23,5 à 90 | 270 à 900 | 1 à 3,5 | |
| **Ortie** | | | | |
| Urtica Doïca | 87 ± 28 | 2,11 ± 0,81 | 1 594 ± 640 | |
| **Ramie** | | | | |
| General data | 24,5 à 128 | 1,2 à 4 | 400 à 1 000 | |
| **Jute** | 26,5 | 1,5 à 1,8 | 393 à 773 | 1,44 |
| **Sisal** | | | | |
| General data | 9 à 21 | 3 à 7 | 350 à 700 | 1,45 |
| **Coir** (coconut) | 4 à 6 | 15 à 40 | 131 à 175 | 1,15 |
| **Cotton** | 5,5 à 12,6 | 7 à 8 | 287 à 597 | 1,5 à 1,6 |
| | | **Animal fibres** | | |
| **Silkworm** | | | | |
| Attacus atlas. | 5 | 18 | 200 | |
| Bombyx mori | 16 | 15 | 650 | |
| **Spider** | | | | |
| Argiope trifasciata | 7 | 30 | 600 | |

### *2.2. Experimental procedure*

#### *2.2.1. Tensile test*

The experimental program was carried out on a hydraulic tensile machine with a capacity of 250kN with a displacement speed of 0.25mm/s during the loading until the failure of specimens. The mechanical behaviour was measured through a tensile test on samples specimens. Figure 1 shows the testing machine. The length of all specimens was 20cm and tested five samples per each fibre fabrics. The ultimate tensile strength of each sample was determined.

**Figure 1**. Experimental tensile test set-up.

*2.2.1. Flexural test*
The parallelepipedal concrete test-tubes are tested in four points bending where the influence of the shear strength is null between the two load points. The loading case was the crossmember speed of 0.5mm/s in a scale of ±50mm with a maximum load cell of 250kN. The specimens were 100x100x400mm sizes.

**Figure 2**. Experimental bending test set-up.

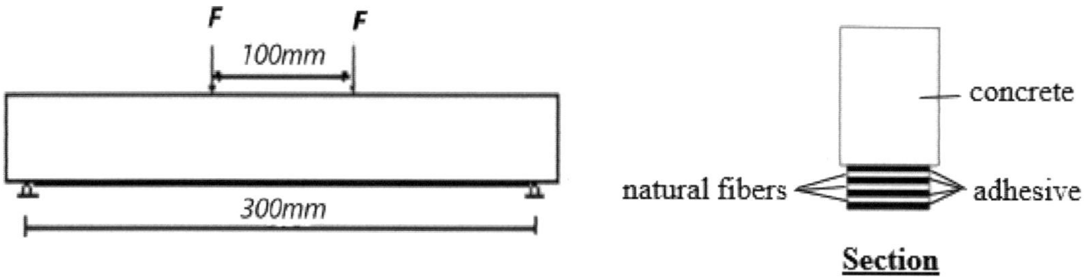


**Figure 3**. Strengthened structure.

## 3. Results and discussion
Mechanical behaviour of composites depends on the individual properties of each component. However, bonding between fibres and matrix plays a predominant role in the material performance. The adequate mechanical behaviour of composite materials is provided with a well-established interface linkage. This zone of property gradients determines the stress transfer between the bonded fibres and the matrix.

*3.1. Influence of different type of natural fibre fabrics*
Five fibre fabrics were tested: fabric of hemp, flax, glass, carbon and mixed hemp+cotton. Table2 and Figure 4 shows the effect of type of fibre fabrics on the ultimate load of composite material sample in tensile test. $F_{u\_hemp}$ represents the ultimate failure load of the hemp fibre fabric.

**Table 2.** Ultimate load of different fibre fabrics.

| | Hemp | Flax | Carbone | Glass | Hemp+cotton |
|---|---|---|---|---|---|
| Fu [kN] | 0.25 | 0.38 | 0.95 | 0.57 | 0.74 |
| Fui/Fu_hemp | 1 | 1.52 | 3.80 | 2.28 | 2.96 |

Carbon fibre fabrics show good mechanical tensile strength almost three times that of hemp fibre fabrics. Knowing that in the case of bending reinforcement of concrete beams, only one third of the capacities of carbon fibres are used, the results displayed by natural fibres (Hemp, linen, and the mixed hemp/cotton) in tensile tests are encouraging for them used in reinforcement.

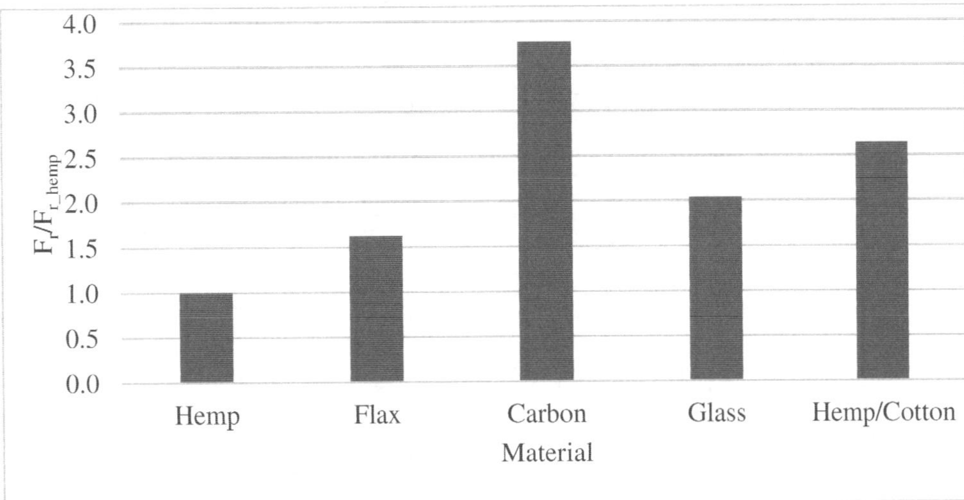


**Figure 4**. Experimental tensile test set-up.

However, strengthening by carbon and glass fibre fabrics are known and validated by recommendation documents of AFGC for engineers, for example. The results in Figure 4 show that the mixing of fibres fabrics could be better than the resistance to the mechanical strength of the strengthened structures tested.

We compare usefully, with the same type of fibre fabrics (one layer of thickness) on the concrete beams in bending, Table 3 and Figure 5.

**Table 3**. Effect of ultimate load of different fiber fabrics.

| | **Control specimen** | **Hemp FCP** | **Flax FCP** | **Glass FCP** | **Hemp+cotton FCP** | **Carbone FCP** |
|---|---|---|---|---|---|---|
| Fu [kN] | 12.09 | 13.02 | 15.74 | 15.72 | 16.32 | 23.5 |
| $F_u/F_0$ [%] | 0% | 8% | 30% | 30% | 35% | 94% |

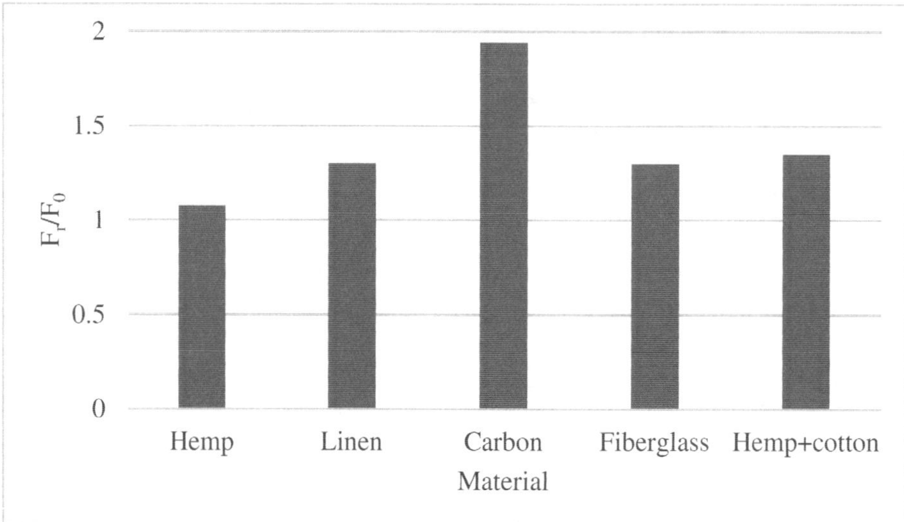


**Figure 5**. Experimental bending test set-up.

$F_0$ is the ultimate load of the concrete beam without strengthening. The mechanical resistance to bending of natural fibres such as hemp, flax, mixed hemp+cotton are outstanding. The results show a benefit in the strengthening of concrete beams by a single layer of each type of composite fibre fabric, in bending tests by increasing the bearing load of structures from 8% to 35%.

### *3.2. Influence of different number of layers*

Table 4 and Table 5 show the effect of each layer fibre fabrics (Hemp and flax) on the ultimate load of strengthened concrete beam. In this part, we are looking for the effect of the layer thicknesses of composite fibre fabrics. As in the previous paragraph, $F_0$ is the ultimate load of the concrete beam without strengthening, control specimen.

Indeed, the results show that from two layers of natural fibre fabrics, the bearing ultimate load of the structure represents more than double of the reference load $F_0$. Epoxy adhesive used in this study, was more malleable fluid, which favoured its penetration into the fibres used. The failures are more cohesive and the behaviour more ductile.

**Table 4**. Influence of hemp fabric layers in the strengthened concrete.

| Hemp | CBH_1 | CBH_2 | CBH_3 |
|---|---|---|---|
| $F_r/F_0$ | 1,12 | 2,50 | 2,20 |

CBH_1: **C**oncrete **B**eam reinforced by **one** layer of **H**emp fabric
CBH_2: **C**oncrete **B**eam reinforced by **two** layers of **H**emp fabric
CBH_3: **C**oncrete **B**eam reinforced by **three** layers of **H**emp fabric

The preceding reports are valid for the two types of fabrics in natural fibres, hemp and linen. In the future, these results are encouraging to continue this research to testing the reinforced concrete beams. The results of the tests on concrete beams reinforced with plates made of flax fibre fabrics are slightly better than those of hemp fibres. However, mixed hemp - cotton fabrics show better results even if not all of them are indicated in this article.

**Table 5**. Influence of flax fabric layers in the strengthened concrete.

| Flax | CBF_1 | CBF_2 | CBF_3 |
|---|---|---|---|

| | | | |
|---|---|---|---|
| $F_r/F_0$ | 2,32 | 2,51 | 2,14 |

CBF_1: Concrete **B**eam reinforced by **one** layer of **F**lax fabric
CBF_2: Concrete **B**eam reinforced by **two** layers of **F**lax fabric
CBF_3: Concrete **B**eam reinforced by **three** layers of **F**lax fabric
$F_0$: Ultimate load of concrete beam without reinforcement
$F_r$: ultimate load of concrete beam reinforced by each fibre fabrics

## 4. Conclusion

The interest of using fibres from natural and renewable resources has risen during the last two decades. The objective of this study was to compare the ultimate load of different composite natural fibre fabric plates and to evaluate the influence on different fabric layers. Five types of fibres (hemp, flax, hemp + cotton, fiberglass and carbon fibre) were carryout in tensile tests. Carbon and glass fibres are used as reference. The tensile tests were performant to evaluate the mechanical proprieties of the composite materials.

For the bending tests, the results show a benefit in the strengthening of concrete beams by a single layer of each type of composite fibre fabric, in bending tests by increasing the bearing load of structures from 10% to 34%.

On the other hand, for the influence of many layers of natural fibres on the beams reinforced in bending, are shown in this paper that the results of fabrics made of hemp and flax fibres. In fact, the results show that from two layers of natural fibre fabrics, the bearing ultimate load of the structure represents more than double of the reference load $F_0$.

The adequate mechanical behaviour of composite materials is provided with a well-established interface linkage. This zone of property gradients determines the stress transfer between the bonded fibres and the matrix.

In the future, these results are encouraging to continue this research to testing the reinforced concrete beams.

**Acknowledgements**
Authors would like to acknowledge and appreciate the support given by University Agency Francophonie, (Franch: Agence Universitaires de la Francophonie -AUF), Fund for Scientific Research (FSR) in Bulgaria through the Programs For Bilateral Cooperation 2018 - University Agency of Francophonie, within the framework of the project Study of biomaterials for strengthening reinforced concrete structures, BioRenfort, Ref.КП-06-франкофония/4, which makes this important and effective research.

Thanks also to the support provided by the Project Ref. 12107 with Scientific Investigation Sector of University of Chemical Technology and Metallurgy (UCTM).

**References**

[1] Assih J 1998 *Contribution à l'étude du Renforcement et de la Réparation de Poutre en Béton Armé par Collage de Plaques Composites en Fibers de Carbone*, Thèse de doctorat, UFR Sciences de Reims
[2] Ivanova I 2013 *Comportement mécanique de console courte en béton armé renforcée ou réparée par collage des matériaux composites*, Thèse de doctorat, Université de Reims Champagne-Ardenne, Reims France
[3] Ivanova I, Assih J, Li A, Dontchev D and Delmas Y 2015 Experimental investigation into strengthened short reinforced concrete corbels by bonding carbon fiber fabrics, *Journal of Adhesion Science and Technology* **29** 2176–89
[4] Ivanova I, Assih J and Dontchev D 2020 Influence of anchorage length of composite fabrics and bonded surface on the strengthened short reinforced concrete corbel by bonding CFRF, *European Journal of Environmental and Civil Engineering* **24** 1993-2009
[5] Luna1P, Lizarazo-Marriaga J and Mariño A 2016 Preliminary Study on the Compatibilization

Techniques of Natural Fibers as Reinforcement of Polymeric Matrices, *Fourth International Conference on Sustainable Construction Materials and Technologies SCMT4,* Las Vegas, USA, August 7-11

[6] Ren G, Patel H, Hapuarachchi D, Crowther S, Fan M and Hogg P J 2009 Development of Natural Hemp Fibre Sheet Mould Composites (NF- SMC), *17th International Conference on Composite Materials (ICCM-17),* Edinburgh, United Kingdom

[7] Colclasure B C, Taylor K. Ruth T K, Brooks T D and Holmes A E 2021 Hemp, Hemp, Hooray: The Impact of a Hemp Educational Campaign on College Students' Attitudes and Knowledge of Industrial Hemp, *Journal of Agricultural Education* **62** 246-59

[8] Tofan L, Paduraru C, Volf I and Balan C 2010 Removal of lead (II) from aqueous solution by sorption ofn natural hemp fibers, *Lucrări științifice, seria Agronomie* **53** 150-2

[9] Jariwala H and Jain P 2019 A review on mechanical behavior of natural fiber reinforced polymer composites and its applications, *Journal of Reinforced Plastics and Composites* **38** 441-53

[10] Belay C 2013 Fibres naturelles de renfort pour matériaux composites, *Technique de l'ingenieur, ref. AM5130V2*

IOP Conf. Series: Earth and Environmental Science **960** (2022) 012008 doi:10.1088/1755-1315/960/1/012008

# Study of the anisotropy of carbon steel sheet by the tensile test

**Gina Mihaela Sicoe**

University of Pitesti, Str. Targul din Vale, nr. 110040 Pitesti, Arges, Romania

e-mail: gina_boicea@yahoo.com, gina.sicoe@upit.ro
ORCID: https://orcid.org/0000-0002-5915-3680

**Abstract.** In test laboratories, the characteristics of the sheets shall be determined by the tensile test on specimens taken in the direction perpendicular to the rolling direction, and these characteristics shall be used as the date of entry into the process of calculating the parts made of the sheet. In the case of parts with special characteristics such as safety parts in the automotive industry, or for mechanically stressed parts in a certain direction, it is important to know precisely the values of the mechanical characteristics of the sheet in relation to the rolling direction, so as to ensure the required resistance. Also, in the case of stamped parts, their quality is all the better the lower the anisotropy. The paper presents the results of experimental research on the mechanical characteristics of the sheet DC04 with a thickness of 0,65 mm, determined by tensile testing on sets of 7 test specimens, taken every 15 degrees from the rolling direction.

## 1. Introduction

Metal alloys are polycrystalline aggregates made up of a very large number of small crystals with different orientations [1], [3]. If the grain is uniformly oriented along any grain direction, it has an isotropic behavior. When the orientations of the grains are grouped along some directions, a polycrystalline material with a preferred orientation [11], anisotropic, results.

Anisotropy is a property of the crystalline state [9] and becomes evident in the case of metal processing by cold plastic deformation, rolling, drawing [4]. The anisotropy of metals [2] represents the variation of the properties of the material depending on the direction of orientation [10].

In the case of sheets, the anisotropy represents the variation of the properties of the material in the sheet plane in relation to the rolling direction [5].

The degree of anisotropy is expressed by the plastic anisotropy coefficient r, which represents the ratio between the natural specific deformation in the direction of the width of the specimen and the specific natural deformation in the direction of the thickness of the specimen. In test laboratories, the characteristics of the sheets shall be determined by tensile testing [8] on specimens taken in a direction perpendicular to the rolling direction, and these characteristics shall be used as input data in the process of calculating the parts made of sheet metal.

The aim of this research is to accurately determine the values of the mechanical characteristics of the sheet in the sheet plane, so that designers, using these results, optimize the design of parts with special characteristics, such as safety parts in the automotive industry or mechanically stressed parts in certain directions. Also, the results can be used for the optimal positioning of semi-finished sheet metal products in the case of stamped parts, [2] to ensure the best possible quality.

## 2. Experimental determinations

The mechanical properties [6], [8] were determined by the tensile test consisting in the application of a tensile force to a tensile specimen, generally until rupture. To define the behavior of the material, the stress-strain conventional curve was drawn, which is obtained by representing in rectangular coordinates the current tension in the tensile specimen (s), depending on the specific elongation, (ε).

$$\sigma=\frac{F}{S} \quad [N/mm^2] \qquad \qquad \varepsilon=\frac{\Delta L}{L}.100 \ [\%] \qquad (1)$$

Where:
- F is the tensile force;
- S is the cross-sectional area of the tensile specimen;
- ΔL is the linear deformation of an element;
- L is the initial length of an element.

### *2.1. Equipment and devices used*

The ZWICK test machine is equipped with a power sensor, an extensometer and a computer for controlling the machine, recording the characteristic curve and calculating the characteristics. The machine is precision class 1. The maximum value of the load can usually be: 200 kN.

The role of the specimen clamps is to rigidly fix the specimen throughout the test and may take different forms depending on the clamping ends of the specimen. Special jars were used for sheet metal specimens, the tightening being provided hydraulically.

### *2.2. Procedure*

The test was performed under ambient conditions at 23°C. The speed of the machine was chosen depending on the nature of the material:
- in the elastic range and up to the upper flow limit 30 [$N/mm^2s^{-1}$];
- in the plastic field 0.008 / s.

### *2.3. Materials*

DC04 sheet with a thickness of 0.65 mm was used. In table 1 are presented the mechanical characteristics and chemical composition according to [7] EN 10130 - Cold rolled low carbon steel flat products for cold forming - Technical delivery conditions.

**Table 1.** The mechanical characteristics and chemical composition.

| Material | Rp 0,2 [$N/mm^2$] max | Rm [$N/mm^2$] | A [%] min | $r_{90}$ min | $n_{90}$ min | C [%] max | S [%] max | P [%] max | Mn [%] max |
|---|---|---|---|---|---|---|---|---|---|
| DC04 | 210 | 270-350 | 38 | 1,6 | 0,18 | 0,08 | 0,03 | 0,03 | 0,40 |

Samplings was performed according to figure 1.

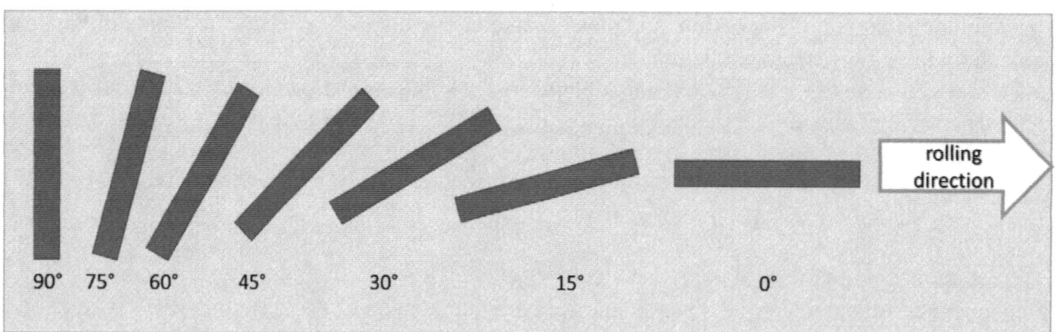


**Figure 1.** Samplings.

For the tensile test, the test specimens were processed according to figure 2. The specimens were obtained by machining.

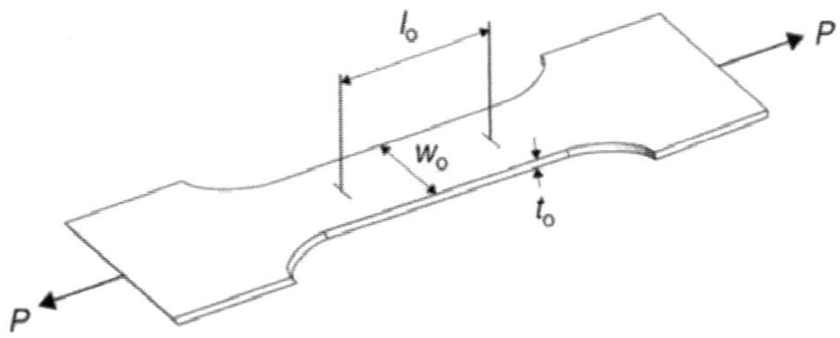


**Figure 2.** Traction test specimen.

The main dimensions are:
- the width of the calibrated area $W_0$ = 20 mm;
- the length of the calibrated $l_0$ = 80 mm;

The characteristics studied in this paper are:
- The conventional yield strength (Rp) represents stress at which the plastic extension is equal to a specified percentage of the extensometer gauge. The symbol used is followed by an index representing the prescribed percentage, e.g. : $R_{p0,2}$.
- Mechanical strength is the property of metallic materials to resist deformation and rupture under the action of internal or external stresses (forces). The breaking strength is the most important property of mechanical materials, being defined as the stress that corresponds to the maximum value of the load (force) for which the deformation capacity is exhausted and the materials break and is given by the relation:

$$R_m = F_{max} / S_0,\ [N/mm^2] \tag{2}$$

where $F_{max}$ is the maximum force and $S_0$ is the initial section of the specimen.
- The plastic anisotropy coefficient (r) is defined as the ratio between the real deformations in width and thickness of a specimen subjected to mono axial traction:

$$r = \frac{\varepsilon_b}{\varepsilon_a} \tag{3}$$

Where:
- $\varepsilon_a$ represents the specific natural deformation in the direction of the specimen thickness;
- $\varepsilon_b$ represents the specific natural deformation in the direction of the width of the specimen.

There were performed 2 series of tests on 7 tensile specimens. The following diagrams, figure 3, show the imagine of the stress-strain conventional curves for all the specimens.

In table 2 are presented the mechanical characteristics of the 2 series of specimens obtained after the tensile test. It is observed the variation of these characteristics depending on the sampling angle of the specimens in the plane of the sheet. This variation is similar in the case of the two series of sheets tested.

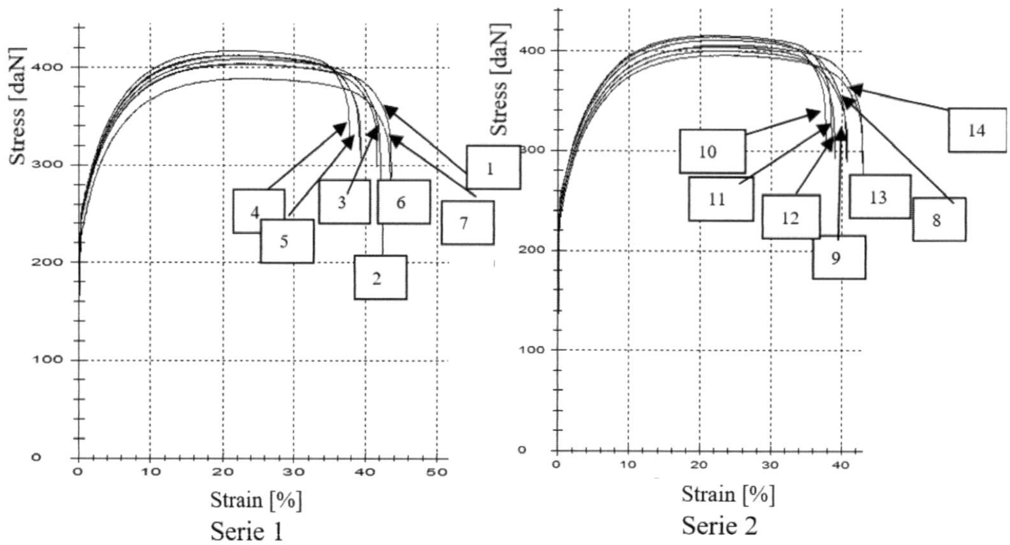


**Figure 3.** Stress-strain conventional curve.

**Table 2.** The results.

| Serie | No. | Sampling angle [°] | $R_{p0,2}$ [N/mm²] | $R_m$ [N/mm²] | $A_{80}$ % | n-value | r-value | E-Modulus [MPa] | m | $A_g$ % | $A_{gt}$ % |
|---|---|---|---|---|---|---|---|---|---|---|---|
| 1 | 1 | 0 | 186,64 | 308,79 | 43,48 | 0,21 | 1,845 | 194.00 | 1,61 | 22,93 | 23,08 |
| 1 | 2 | 15 | 185,21 | 309,19 | 41,52 | 0,21 | 1,620 | 196.55 | 2,04 | 22,16 | 22,31 |
| 1 | 3 | 30 | 187,59 | 315,35 | 39,24 | 0,21 | 1,270 | 197.70 | 1,85 | 21,44 | 21,59 |
| 1 | 4 | 45 | 190,12 | 319,20 | 37,62 | 0,21 | 1,175 | 199.16 | 2,00 | 20,87 | 21,03 |
| 1 | 5 | 60 | 186,16 | 315,18 | 39,03 | 0,21 | 1,290 | 194.27 | 1,89 | 22 | 22,15 |
| 1 | 6 | 75 | 184,47 | 312,63 | 42,36 | 0,21 | 1,500 | 187.14 | 1,84 | 22,33 | 22,48 |
| 1 | 7 | 90 | 178,29 | 306,74 | 43,31 | 0,21 | 1,580 | 182.63 | 1,73 | 23,18 | 23,34 |
| 2 | 8 | 0 | 185,93 | 305,86 | 40,68 | 0,21 | 1,810 | 188.91 | 1,67 | 23,30 | 23,45 |
| 2 | 9 | 15 | 192,96 | 313,74 | 40,43 | 0,21 | 1,680 | 174.60 | 1,50 | 22,87 | 23,07 |
| 2 | 10 | 30 | 190,59 | 316,35 | 37,62 | 0,21 | 1,355 | 190.10 | 1,52 | 21,23 | 21,40 |
| 2 | 11 | 45 | 188,87 | 317,30 | 38,44 | 0,21 | 1,150 | 198.61 | 1,47 | 22,24 | 22,40 |
| 2 | 12 | 60 | 181,29 | 309,07 | 38,91 | 0,21 | 1,300 | 190.08 | 1,55 | 22,30 | 22,46 |
| 2 | 13 | 75 | 176,23 | 302,44 | 42,77 | 0,21 | 1,525 | 181.18 | 1,72 | 22,87 | 23,03 |
| 2 | 14 | 90 | 181,48 | 310,11 | 42,86 | 0,21 | 1,715 | 182.04 | 1,68 | 23,40 | 23,55 |

- $R_p$ - yield strength;
- $R_m$ - tensile strength;
- $A_{80}$ % - percentage elongation after fracture;
- n-value – strain hardening exponent;
- r-value – plastic strain ratio;
- E-Modulus - modulus of elasticity;
- m - slope of the stress-percentage extension curve at a given moment of the test;
- $A_g$ - percentage plastic extension at maximum force, $F_m$;
- $A_{gt}$ -percentage total extension at maximum force, $F_m$;

IOP Conf. Series: Earth and Environmental Science **960** (2022) 012008 doi:10.1088/1755-1315/960/1/012008

## 3. Interpretation of results

For each series of tests, diagrams of variation of the studied parameters were drawn according to the sampling angle of the specimens.

Figure 4 shows the diagram of the variation of the plastic anisotropy coefficient depending on the sampling angle of the specimens for both series of determinations.

It can be observed that the plastic anisotropy coefficient has the maximum value in the rolling direction, shows a significant decrease up to 45° from the rolling direction and increases again in directions perpendicular to the rolling direction to a slightly lower value than in the rolling direction. These results are correlated with hardening of the material which is maximum on the rolling direction.

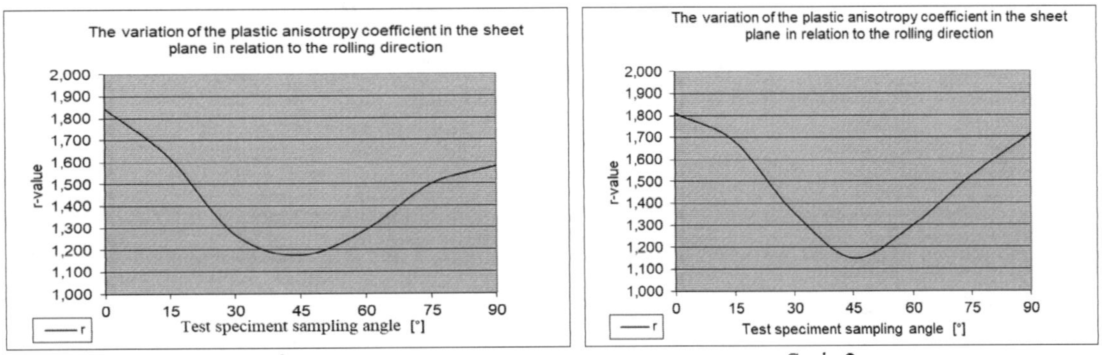


Serie 1 Serie 2

**Figure 4.** Variation of the plastic anisotropy coefficient depending on the sampling angle of the specimens for both series.

In figure 5 are presented the diagrams of the elastic limit and the breaking strength in the sheet plane depending on the sampling angle of the specimens for both series of determinations. These characteristics have very small variations depending on the orientation of the specimens. We can consider that these characteristics are constant in the plane of the sheet.

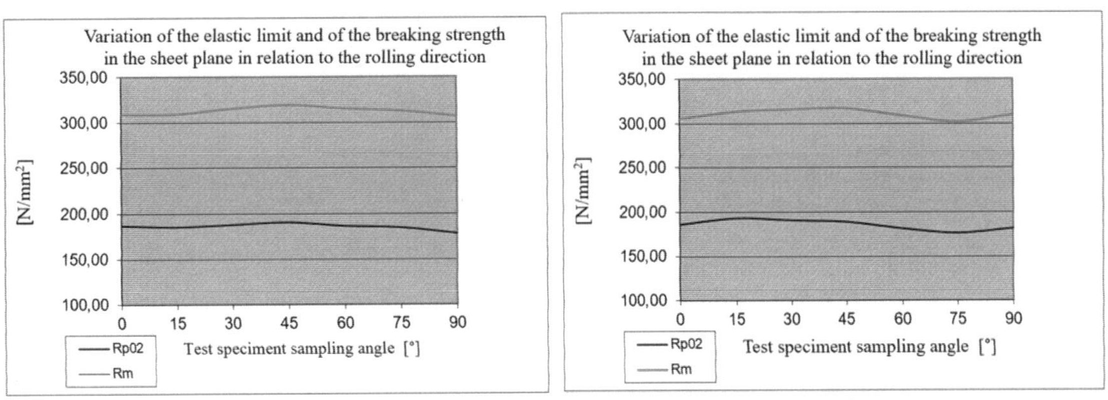


Serie 1 Serie 2

**Figure 5.** The diagrams of the elastic limit depending on the sampling angle of the specimens for both series

Figure 6 shows the diagram of the variation of elongation depending on the sampling angle of the specimens for both series of determinations. Elongation after fracture presents identical values in longitudinal and perpendicular plane and a slight decrease in the direction of 45°. The plastic extension at maximum force can be considered constant in the plane of the sheet.

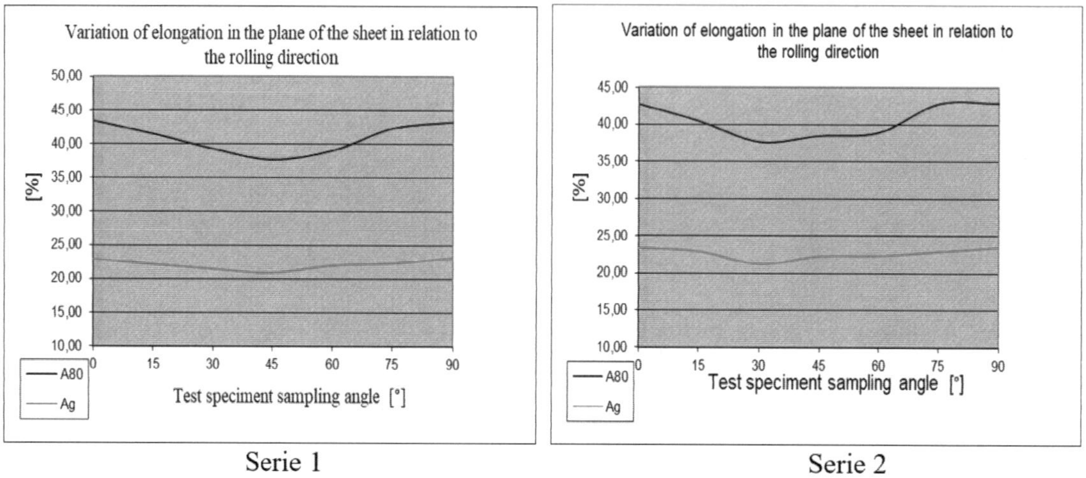


**Figure 6**. The diagram of the variation of elongation depending on the sampling angle of the specimens for both series

## 4. Conclusions

The results obtained on the both series of tests highlighted the variation of the mechanical characteristics in the plane of the sheet DC05 depending on the rolling direction.

It is observed that in the case of cold rolled sheets there is an important variation of the values of the anisotropy coefficient in the sheet plane, in relation to the rolling direction. The plastic anisotropy coefficient has the maximum value in the rolling direction, shows a significant decrease of 36% from the rolling direction and increases again in directions perpendicular to the rolling direction to a slightly lower value than in the rolling direction. These results are correlated with hardening of the material which is maximum on the rolling direction and also are similar for the both series of tests.

In industrial technological processes the anisotropy coefficient is of particular interest, because the behavior of the sheet at stamping is influenced by anisotropy. The so-called drawing ears will appear in the directions in which the material has a higher anisotropy coefficient.

The yield strength and tensile strength have relatively small variations in sheet metal of 2-4%, which can be considered insignificant for the calculation of strength in the case of parts subjected to mechanical stress.

Elongation after fracture presents identical values in longitudinal and perpendicular plane and a slight decrease by 10 to 13%, in the direction of 45°. The plastic extension at maximum force can be considered constant in the plane of the sheet.

Some areas suitable for further investigation may be:

- anisotropy study for high strength steel sheets for automotive applications;
- development of finite element methods for the study of cold rolled sheet anisotropy, depending on sheet thickness and quality.

## 5. References

[1] Tomasz T and Hirpa G, 2011, Investigation of anisotropy problems in sheet metal forming using finite element method, *International Journal of Material Forming*, DOI: 10.1007/s12289-010-0994-7

[2] Bîrlan A and Jădățean M, 2011, Experimental study on the rise ears cylindrical hollow parts, a XI-a *Conferinta Nationala multidisciplinara - cu participare internationala "Profesorul Dorin PAVEL"*

[3] Khelifa M and Oudjene M, 2008, Numerical damage prediction in deep- drawing of sheet metals, J *Mat Proc Technol* **200**:71–76

[4] Yi S, Bohlen J, Heinemann F and Letzig D, 2010, Mechanical anisotropy and deep drawing behavior

of AZ31 and ZE10 magnesium alloy sheets, *Acta Mater* **58**:592–605
[5] Cheng P and Yao Y L, 2005, The Influence of Sheet Metal Anisotropy on Laser Forming Process. *ASME* Vol.**127**: 572-582
[6] Teaca M, Martiny M, Charpentier I and Ferron G, 2009, Identification du comportement plastique des tôles métalliques anisotropes; *9ème Congrès Français de Mécanique Marseille.*
[7] EN 10130 - Cold rolled low carbon steel flat products for cold forming - Technical delivery conditions
[8] ISO 6892-1:2016 - Metallic materials. Tensile Testing Part 1: Method of test at room temperature
[9] B. Hutchinson, 2015, Critical Assessment 16: Anisotropy in metals, *Materials Science and Technology*, **31:12**, 1393-1401
[10] Iordache M, Teaca M, Charpentier I, Martiny M and Ferron G, 2009, Identification of sheet metal plastic anisotropy, and optimization of initial blank shape in deep drawing, *The annals of "Dunărea de Jos" University of Galați,* fascicle **V**, 149-154
[11] Hutchinson B, Ryde L and Bate P, 2005, Transformation textures in steels', *Mater. Sci. Forum*, 495-497

# Comparative modal analysis for a twisted beam axle for a hybrid vehicle

**Laurian Tomşeanu, Viorel Nicolae, Ionel Vieru, Dănuţ Marinescu**

University of Piteşti, Faculty of Mechanics and Technology, Târgu din Vale Street, no.1, Piteşti, Romania

E-mail: viorel.nicolae@upit.ro ORCID ID : 0000-0001-7598-7434

**Abstract**. In this article it is presented a comparative analysis of natural mode frequencies for a non-powered rear axle used to fit mass production vehicle and for a similar rear axle derived from the first one to be used to fit a hybrid powered vehicle. The CAD model of the axle and the computed natural mode frequencies were realised using CATIA V5. For calculation, finit element method was used.

## 1. Introduction

Modal analysis consists into finding the dynamic characteristics of the structure, and it might include both analytic and experimental processes [1].

Deploying the modal analysis, we assume that the part vibrates under the action of an initial impulse; in this case, the natural mode frequencies and vibration frequencies represent interesting issues, so the natural mode frequencies represent one of the multiples ways a part is deforming when vibrating [2].

The advantages of using the modal analysis when study the system consist in estimation of the dynamic characteristic of the structure and in determination of eigen frequencies.

Once determinate, the eigen frequencies can be used for:

- frequencies analysis and transitory analysis;
- experimental analysis;
- optimization of structure dynamic behavior.

Using a rear non powered axle for a mass production vehicle, by making simple modification it was converted into a a rear axle which can be used on an hybrid vehicle.

The proposed solution for transforming the rear axle into an electric powered one it is simple: the electric engine is suspended on the car body, the transmission operates through a bevel gears directly fitted to the differential as shown in Figure 1.

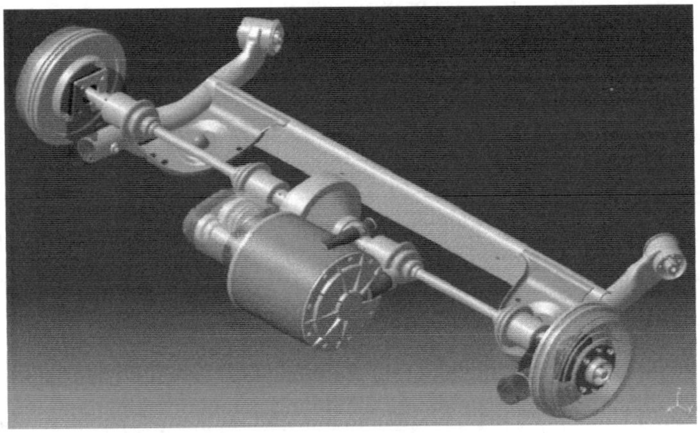

**Figure 1.** Rear electric powered axle.

The main changes needed to transform the non-powered axle (further named SERIAL) into a powered axle (named HYBRID) were made mostly in the area where wheels are mounted, as shown in Figure 2.

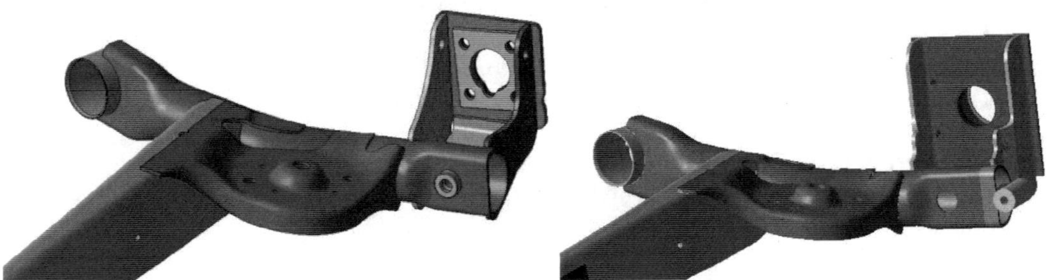

a) SERIAL axle b) HYBRID axle

**Figure 2.** The differences between SERIAL axle and HYBRID axle.

The changes consist in length extension of the arms with 20 mm and the shock absorber plate was moved to the end of the arm (to allow mounting of the shafts). The spindle holder was modified to allow fitting the new spindle, and also to allow mounting the new shafts to be connected with the wheel hub.

## 2. Finite element method used in calculation of modal analysis

Mainly there can be done two types of modal analysis:

- free - free (free frequency analysis), when the constraints are not taken into account (border constraints or limit conditions);
- with enforced condition (frequency analysis), when the constraints are taken into account.

During the “free – free” modal analysis, the structure is considered suspended and so it is not applied any stress or constraint. This type of analysis allows to obtain pure characteristics of a structure, without influences imposed by constrains or exterior environment (forces, moments, and so on).

A modal analysis can be made, also, with particular limit conditions; this constraints are, in the majority of the cases, those encountered in usage .

The results of the modal analysis are resonance frequencies, vibration modes, the deformed structure or the deformed assembly determined by the resonance frequencies and the way the vibration modes influence the vibration of the structure or of the assembly on a specific direction. Because the vibrational characteristics of the structure determine their response to any type of dynamic load, the modal analysis is done before any dynamic analysis [3].

Generally, within the structural analysis programs, the modal analysis is an linear analysis. Any non-linearities are ignored. Neglecting the damping effect, the free vibrations are described by the following equation:

$$[M]\{\ddot{u}\} + [K]\{u\} = \{0\} \tag{1}$$

where:
[M] – inertia matrix or the matrix of masses;
[K] – rigidity matrix;
{u} - vector of nodal displacements ;
$\{\ddot{u}\}$ – vector of nodal accelerations.

Matrices [M] et [K] are global for the entire structure, being obtained by assembly the submatrices corresponding to each finit element.

The second order differential equation system (1) is a unitary system with constant coefficients. Taking into account the physical nature of the phenomenon, it is assumed that the solution of the equation (1) is like :

$$\{u_i\} = \{\Phi\}_i \cos \omega_i t \tag{2}$$

In which $\{\mathbf{\Phi}\}_i$ – eigen vector representing the deformed structure, or the eigen vibration mode corresponding to its eigen pulse $\omega_i$ [rad/s], or eigen frequency $f_i$ [$s^{-1}$].

During the modal analysis, the primary unknowns of the equation are the eigen vectors $\{\Phi\}_i$. They contain the relative amplitudes, or the distribution coefficients of the vibration modes corresponding to their eigen frequency $f_i$, or to their resonance frequencies of *i* rank.

In order to be determined, the eigen vector is normalized.

Using the solution (2), the equations system (1) is rewrite in the form of a problem with eigenvectors and eigenvalues:

$$(-\omega_i^2 [M] + [K]) \{\Phi\}_i = \{0\} \tag{3}$$

Eigen frequencies are calculated with the relation:

$$f_i = \frac{\omega_i}{2\pi}; \tag{4}$$

## 3. The calculation models

The discretization must be fine enough so the modal deformation to be correctly represented, being necessary to be at least four elements on the wavelength.

This involve achieving of a fine discretization by growing the frequency, because the wavelength is decreasing when the frequency increase.

During the study of the frequency bands of the axle elements, the discretization level used for the dynamic study is generally the same with the one used for static studies.

Usually when finite element modeling is carried out, it is not only to reflect the mechanical properties of the structure, but also to adopt the appropriate unit configuration and the number of elements as few as possible. Before dividing the mesh, we should divide the plate parts of different thickness into different parts, which can be divided into the longitudinal swing arm, the torsion bar, the frame and so on (Fig. 2).

In order to perform this calculation, the axles were discretized using tetrahedron elements with 10 [mm] medium side length, as presented in Figure 3.

**Figure 3.** HYBRID axle discretization into finite elements

The calculation was made using a FFA (Free Frequency Analysis) from Generative Structural Analysis module of CATIA V5 application [4].

The material properties used in the calculation are:

- elasticity module (Young's modulus) E = 2,1*10 5 MPa
- Poison's ratio n = 0,300
- material density $\rho = 7800$ kg/m$^3$

The damping phenomenon of the structure is not taken into account in the modal analysis calculation for a damping lower than 15% (the majority of the studied cases). In this case, the eigen frequencies were very little impacted.

## 4. The results obtained

Table 1 presents the first nine eigen vibration modes calculated with the correspondent values of frequencies and the deformation mode of the axles for each individual case (reference system Fig. 3) .

**Table 1.** Eigen vibration modes calculated for two axles.

| No. of eigen mode | SERIES AXLE | | HYBRID AXLE | |
|---|---|---|---|---|
| | Freq. [Hz] | Modal deformations | Freq. [Hz] | Modal deformations |
| 1 | 48 | Axle torsion around OY axis | 41 | Axle torsion around OY axis |
| 2 | 104 | Axle bending on XOY, in fase with bending of anti-roll bar | 96 | Axle bending on XOY, in fase with bending of anti-roll bar |
| 3 | 119 | Axle bending on YOZ, in fase with bending of anti-roll bar | 114 | Axle bending on YOZ, in fase with bending of anti-roll bar |
| 4 | 121 | Axle bending on XOY, in antiphase with bending of anti-roll bar | 121 | Axle bending on XOY, in antiphase with bending of anti-roll bar |
| 5 | 189 | Axle bending on YOZ, in antiphase with bending of anti-roll bar | 183 | Torsion of anti-roll bar and the connecting element around Y axis |
| 6 | 269 | Axle torsion over OZ axis | 239 | Axle torsion over OZ axis |
| 7 | 302 | Bending of the linked elements on XOY | 276 | Bending of the linked elements on XOY |
| 8 | 310 | Axle bending on XOY in antiphase with bending of the anti-roll bar with two curves | 303 | Axle bending on YOZ in antiphase with bending of the anti-roll bar with two curves |
| 9 | 319 | Axle bending on YOZ in antiphase with bending of the anti-roll bar with two curves | 317 | Axle bending on XOY in antiphase with bending of the anti-roll bar with two curves |

Table 2 presents the images of the corresponding shapes for each eigen vibration mode calculated before.

Analyzing the information presented in Table 1 and Table 2, it could be noticed that the values of the eigen frequencies corresponding to each eigen mode are smaller for the HYBRID axle compared with the values of the SERIES axle.

Also, the shape of vibration mode are similar for both axles excepting the eigen modes 8 and 9.

**Table 2**. Images with the shapes of the eigen vibration mode for the axles analyzed.

| No. of eigen mode | **SERIES AXLE** | **HYBRID AXLE** |
|---|---|---|
| 1 | 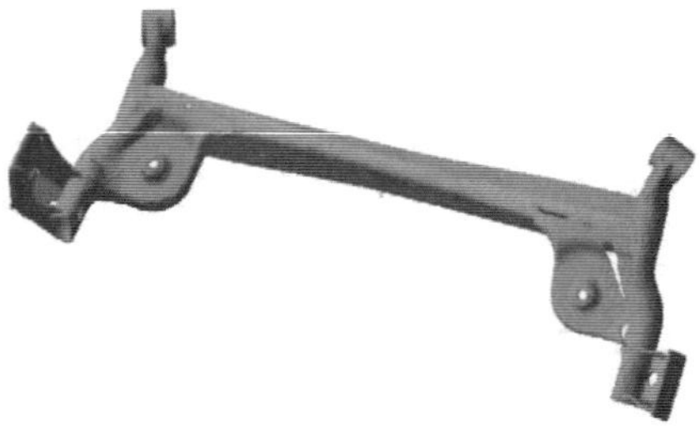 |  |
| 2 | 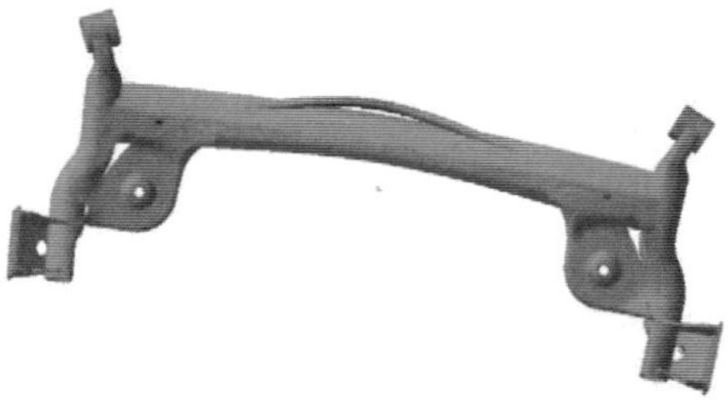 | 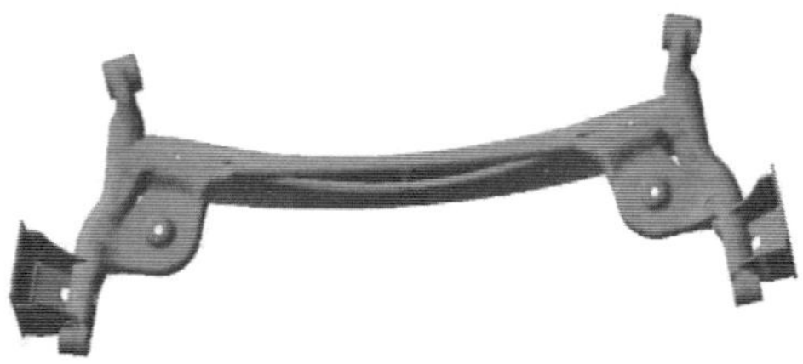 |
| 3 | 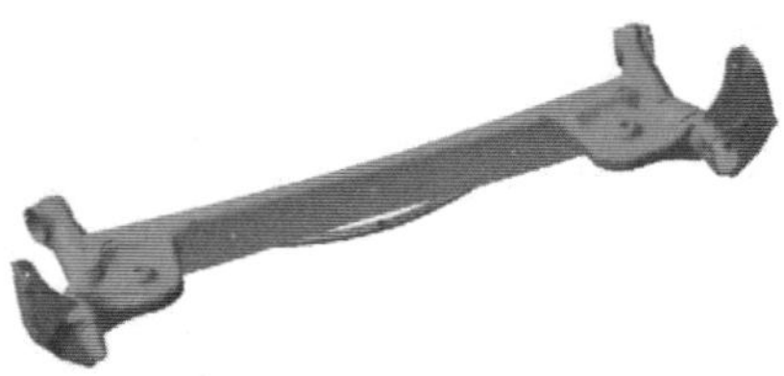 | 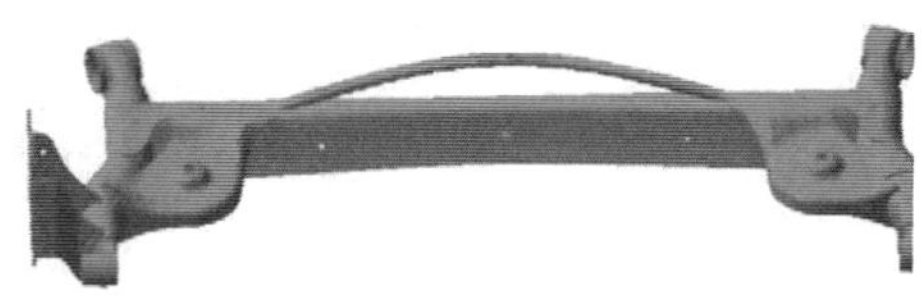 |
| 4 |  | 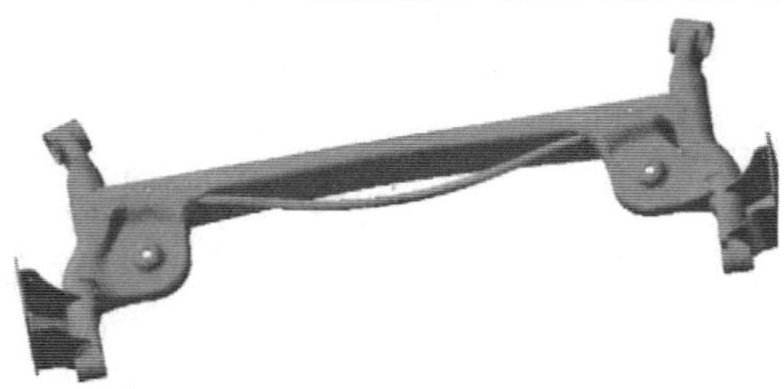 |

| | | |
|---|---|---|
| 5 | 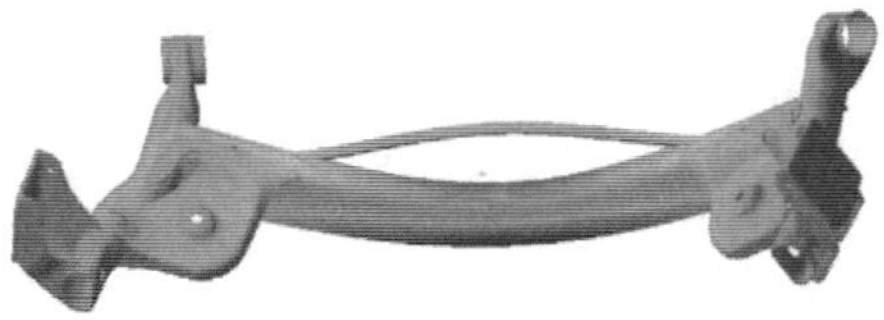 |  |
| 6 | 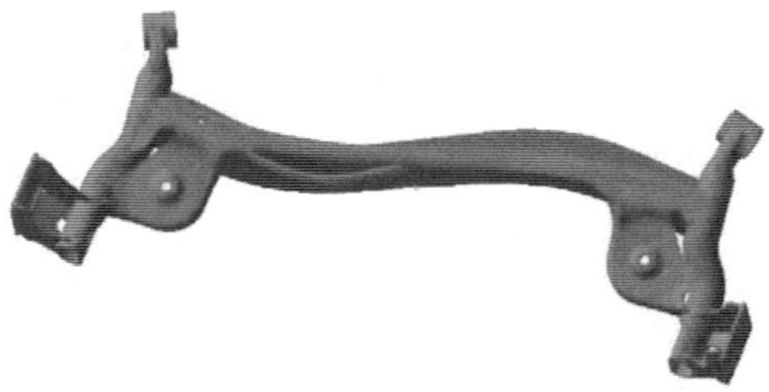 | 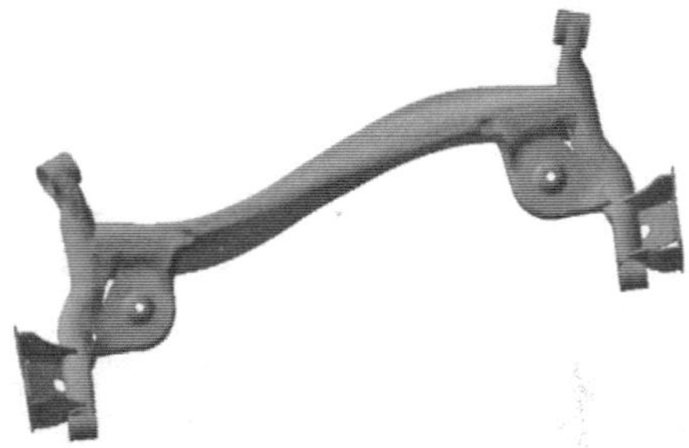 |
| 7 |  | 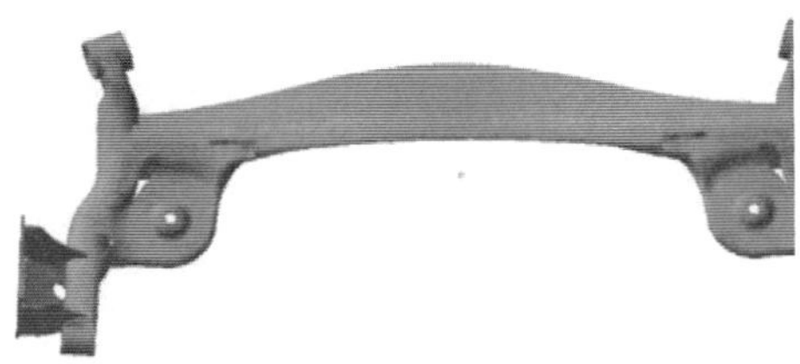 |
| 8 | 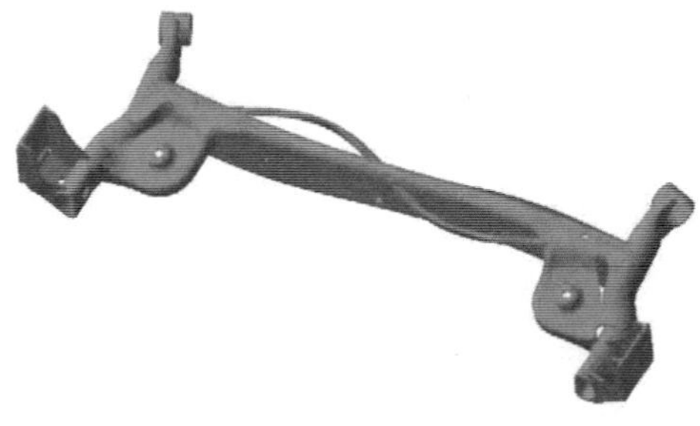 | 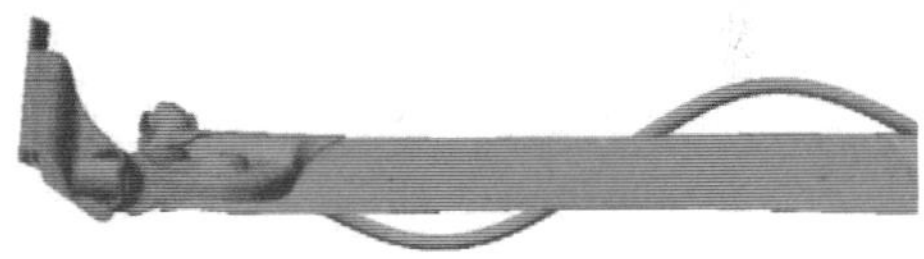 |
| 9 |  | 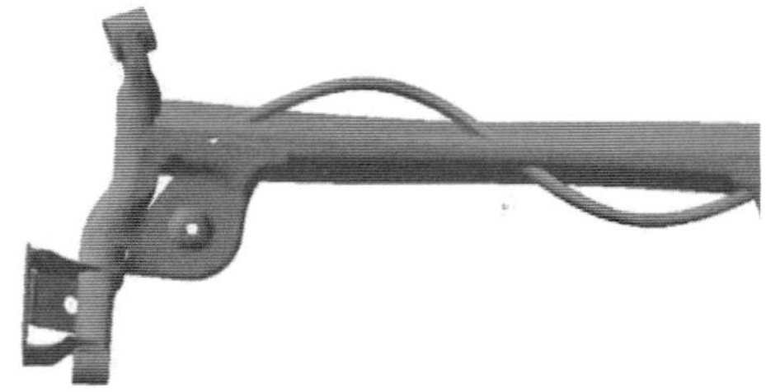 |

## 5. Conclusions

Complex assemblies are impossible to be entirely modeled due to the many hypothesis made to simplify the simulation process. In many situations it is used the experimental validation because it leads to additional confidence that the models are giving precise results. Also, the experimental results can be used to represent the components that can't be precisely designed and to establish the internal dumping degree of the structure.

However, knowing the eigen frequencies of the axles allow to do a better selection of the frequency interval studied for other dynamics calculations, e.g. the calculation of the frequency response. Also, from the experimental point of view, is interesting to realize a modal numerical analysis before the experiment in order to know the strains and the eigen frequencies and so to determine the correct position for the placement of the accelerometers.

Further, the modal base of the axle assembly could be compared with the modal base of the car body or of some elements from near area and so to avoid the common or nearness frequencies by redirecting the design to certain frequencies.

## References

[1] Ewins D J 2000 *Modal Testing Theory, Practice and Application* (Research Studies Press Ltd.)
[2] Zhi-Fang Fu, Jimin He 2001 *Modal Analysis* ISBN 0 7506 5079 (Butterworth-Heinemann)
[3] Cook R D, Malkus D S, Plesha M E and Witt R J 2002 *Concepts and Applications of Finite Element Analysis* (John Wiley & Sons, USA)
[4] *** 2003-2009 *Documentation CATIA version 5* Release 10-19 (Dasaults Systems)

IOP Conf. Series: Earth and Environmental Science **960** (2022) 012010 doi:10.1088/1755-1315/960/1/012010

# Is it possible to turn a diesel engine into a less pollutant device?

**Cosmin Constantin Suciu**[1[0000-0002-8754-534X]], **Ioana Ionel**[1[0000-0002-2999-1210]], **Daniel Ostoia**[1 [0000-0002-6054-7062]], **Nicolae Stelian Lontis**[1[0000-0002-6072-4399]], **Ion Vetres**[1[0000-0002-6425-9818]], and **Sorin Vlad Igreț**[2]

[1]PolitehnicaUniversity Timisoara, Faculty of Mechanical Engineering, Timisoara, Romania
[2]Aurel Vlaicu University, Faculty of Engineering, Arad, Romania

ioana.ionel@upt.ro

**Abstract**. The current tendency in the eastern part of Europe is to modify old diesel engines with the purpose of improving characteristics in terms of horsepower and torque, but also to reduce the generated pollution. The diesel engines are still in use, against to the tendencies of renouncing to their support, at least in larger urban & industrial areas, where the pollution level, especially the particulate matter (particles PM10, PM2.5) and ozone concentrations, also the $NO_x$ are supposed to be generated mostly by diesel engine vehicles. The paper presents results concerning the influence of modifying the diesel engine control unit's parameters, such as injection quantity, start of injection, intake air pressure and all the others correlated for better performance. The article brings into attention possibilities to reduce the exhaust pollution concentrations, correlated by simultaneous ways of improving the external characteristics of the engine by modifying the engine control unit's parameters. Measurement results of a compression ignited internal combustion engine before and after the retrofitting, by reparametrization, meaning changing the parameters are presented and discussed.

## 1. Introduction

Is it possible to turn the diesel into green energy [1]? This article is continuing the studies concerning the impact on the pollution and climate change generated by the vehicles [2, 3]. After describing the impact on the environment and organisms of the pollutants produced by internal combustion engine, this article focuses on how to reduce them, but also on the improvement of a compression ignited engine's emissions, efficiency and performance characteristics by remapping its control unit. The current tendency in the Eastern part of Europe is to modify old diesel engines with the purpose of improving characteristics as horsepower and torque. It might be a transitorily solution. The experiment presents also the measurement results of a compression ignited internal combustion engine, before and after the reparametrization (retrofitting the parameters) of its control unit, with the purpose of not only informing the reader of the benefits but also finding ways on obtaining green energy from it.

As a continuation of a previous article [4], one is analyzing the outcome in terms of engine's emission values, characterized by factory parameters, in comparison with the results obtained after a reparameterization of the engine control unit.

One of the most damaging result in diesel engines are the nitrogen oxides [5]. They are forming acids that, on humans can cause irritation of the bronchial tubes, coughing and people suffering from asthma can be very sensitive to $NO_x$. In terms of the terrestrial and aquatic ecosystems the impact is also negative

[6]. Areas with $NO_x$ emissions contain higher levels photolytic source of OH, stimulating $O_3$ production that leads to O3 rich air and quality problems, while in the ocean areas, OH leads to lower levels of O, with a negative impact on the climate control [7].

Black smoke and black carbon are radiative forcing agents. This particulate matter (PM10) stays in the air for hours or even days. Being bonded to other harmful substances, they enter the respiratory system and into the lungs leading to a big number of health problems [6, 7].

Table 1 presents an overview of the parameters used for the Eco-score methodology, for a better awareness of its impact and why reducing pollution is necessary.

**Table 1.** Eco-score methodology parameters overview [6].

| Classification | Weighting α | Inventory | Units | Characterization Rural | Characterization Urban |
|---|---|---|---|---|---|
| Global warming | 50% | $CO_2$ | GWP | 1 | 1 |
| | | $CH_4$ | GWP | 23 | 23 |
| | | $N_2O$ | GWP | 296 | 296 |
| Air quality | (40%) | | | | |
| Human health | 20% | KWS | €/kg | 3 | 3 |
| | | CO | €/kg | 0.0008 | 0.0032 |
| | | PM10 | €/kg | 103.49 | 418.61 |
| | | $NO_x$ | €/kg | 1.152 | 1.483 |
| | | $SO_2$ | €/kg | 6.267 | 14.788 |
| Ecosystems | 20% | $NO_x$ | €/kg | 0.176 | 0.176 |
| | | $SO_2$ | €/kg | 0.113 | 0.113 |
| Noise | 10% | Sound level | dB(A) | x-40 | |

After consulting the results, it is clear that diesel engines have the highest impact on human health, the Eco-score methodology being mostly influenced by the high number of particle matter (PM10), followed by $NO_x$ and $SO_2$, which influence both human health and ecosystems [6].

Having an overview, one has noticed the higher demand in the Eastern Europe areas in modifying diesel engines, targeting mostly euro 3 due to the easier way of obtaining higher values in both horsepower and torque and not keeping under observation the high level of emissions, soot and black smoke as a result. Since the measuring is done to obtain the difference between a factory specification vehicle with both a factory software and a modified one. For further articles, one will use the current results into reducing the levels of pollutants to lower levels while increasing engine's horsepower and torque characteristics. Furthermore, rising awareness on the negative effects of modifying engine's power characteristics while neglecting pollution and the damage produces by it.

## 2. Engine specifications and fuel system

A4-cylinder, compression ignition 1896 $cm^3$, 2 valves per cylinder, turbocharged with intercooler, traverse alignment engine equipped with, vehicle, is measured.

This has a bore of 79.5 mm with a stroke of 95.5 mm with a compression ratio of 19.5, being equipped with a VP37 Bosch injection pump and a Garrett GT1749V [8] variable geometry turbocharger. It measures an output of 89 bhp at 3750 rpm with 210 Nm of torque at 1900 rpm.

Te direct injection diesel engine is equipped by 4 mechanical injectors, with 0.184 micrometer nozzles, having 5 holes and a 10 mm VE injection pump from Bosch (VP37).

For the two staged injectors, the pre-injection takes place at 220 bar and the second stage (main injection) takes place when the pressure reaches the value of 300 bar.

## 3. Measurements

The measurements were achieved using the "Capelec 3200" stand-alone opacimeter, capable of offering multi-gas analysis and road-worthiness testing. It is characterized by ISO 11614, NFR 10025-2016 and CEM approvals and certifications, meaning that it can also be used for the technical inspection.

The machine is also upgraded with a Capelec gas analyzer module, meaning one can obtain data from both petrol and diesel engines experiments.

Measurements have been achieved in a certified and authorized periodical inspection institution meaning the Romanian Automotive Register, attested through the Product Certification Body (RAR-OCP), with SR EN ISO/CEI 17065:2013 certification.

## 4. Results

**Table 2.** Opacimeter measurement results.

| | Factory software | | Modified software | |
|---|---|---|---|---|
| Engine speed [RPM] | 903 | 3000 | 903 | 3000 |
| K (light absorption coefficient) | 0.033 | 0.120 | 0.029 | 0.196 |
| Engine Oil temperature [°C] | 105 | 105 | 108 | 108 |

From the first measurement with the opacimeter, presented in table 2, one concludes that, at idle speed, after modifying the software, the results show less soot. For example, the value is 13% less, at 3000 rpm. But, one found a value of 63 % more particulate matter, eliminated on the exhaust, in comparison to the initial case.

**Table 3.** Gas analyzer measurement results at engine idle speed.

| | Factory software | Modified software |
|---|---|---|
| Engine speed [RPM] | 903 | 903 |
| CO [% vol.] | 0.01 | 0.01 |
| $CO_2$ [% vol.] | 1.20 | 1.10 |
| $O_2$ [% vol.] | 19.28 | 18.75 |
| HC [ppm vol.] | 0 | 0 |
| $NO_x$ | 0 | 0 |
| Lambda | 12.19 | 12.87 |
| $CO_{corr}$ [% vol.] | 0.12 | 0.14 |
| Engine oil temp. [°C] | 106 | 107 |

The cleaner idle speed status noticed firstly on the opacimeter's measurements is generated from the difference of the stoichiometric ratio, (see table 3), which is higher in this case, by 5% than compared to the factory values; a decrease by 2.8 % of oxygen eliminated on the exhaust is noticed. Knowing that the gas analyzer is used only for petrol engines at the periodical technical inspection, due to the very high stoichiometric ratio on the diesel engine, it shows a "0" value, for both HC and $NO_x$.

**Table 4.** Gas analyzer measurement results at an engine speed of 3000 rpm.

| | Factory software | Modified software |
|---|---|---|
| Engine speed [RPM] | 3000 | 3000 |
| CO [% vol.] | 0.01 | 0.03 |
| $CO_2$ [% vol.] | 1.30 | 1.50 |
| $O_2$ [% vol.] | 19.01 | 18.19 |
| HC [ppm vol.] | 0 | 0 |
| $NO_x$ | 0 | 0 |
| Lambda | 11.19 | 9.35 |
| $CO_{corr}$ [% vol.] | 0.11 | 0.29 |
| Engine oil temp. [°C] | 106 | 105 |

On the other hand, the same pattern as before is to be noticed in table 4. Compared to the idle speed being less polluting, at 3000 RPM, the air excess coefficient is lower by 19.6% and there is an increase in $CO_2$ by 15%.

**Table 5.** Engine's parameters differences for the presented measurements.

| | Factory software | | Modified software | |
|---|---|---|---|---|
| Engine speed [RPM] | 903 | 3000 | 903 | 3000 |
| Intake air pressure [mbar] | 1020 | 1091.4 | 1020 | 1081.2 |
| Injection quantity [mg/str] | 2.8 | 4.4 | 2.6 | 3.5 |
| Start of injection [°BTDC] | 0.4 | 0.2 | 0.7 | 0.2 |
| Coolant temperature [°C] | 90.9 | 91.8 | 90 | 91.8 |

Even if the injected fuel quantity has lowered after the software modifications in both cases, the measurements from table 5 proved different results. Further analyze will be planned, regarding the topic and how also modifying the engine's mechanical components can influence its function. The final target is to obtain a less pollutant diesel engine, with increased performance, and lower fuel consumption, an engine that could be an option, at least in the next years, for the gas ones, on hybrid vehicles. The modification towards the engine control unit will be detailed in a separate article, dedicated specifically to this purpose, and based of course on many parameters and practical interpretation of achieved engine novel characteristics and behavior.

## 5. Conclusion

The paper resumes some opportunities to turn a diesel engine into a more environment friendly engine. Thus, the reparameterization of the electronic control unit has been done together with the cancellation of the exhaust gas recycling valve with the purpose of increasing the engine's power performance, as result one found that the $CO_2$ reached values higher by 15% compared to the initial ones. Even if the injected values are lower and the fuel consumption might improve, the exhaust output is worse than with the factory software, proving that if that target is to increase horsepower, the pollutant level are also increasing, and this is not recommended. The measurements have been realized in a static regime with no load.

## 6. References

[1] Luke Gibson, Elspeth N Wilman, William F Laurance, How Green is 'Green' Energy?, *Trends in Ecology & Evolution,* Volume **32**, Issue 12, Pages 922-935, ISSN 0169-5347,2017.
[2] Giavazzi, Fulvio & Buttini, Patrizia & Perego, Carlo. *Polluting emissions caused by transport.*, pp. 2-33, 2007.
[3] Hester R E and Harrison R M, *Air Quality in Urban Environments*, pp. 86-87, 2009.
[4] Cosmin Constantin Suciu, Daniel Ostoia, Nicolae Stelian Lontis, Ion Vetres, Sorin Vlad Igreț, Ioana Ionel, Preliminary research concerning the pollution generated by a diesel engine, in order to higher its performance, *The 11th International Conference on Machine and Industrial Design in Mechanical Engineering,* accepted for publication, Springer, 2021.
[5] John Pignon, *Diesel Engines: Design and Emissions*, Review of a course on diesel particulates and NOx emissions , pp.119, 2005.
[6] Timmermans, Jean-Marc & Matheys, Julien & Van Mierlo, Joeri & Lataire, Philippe. Environmental rating of vehicles with different fuels and drive trains: A univocal and applicable methodology. *European Journal of Transport and Infrastructure Research.* 2006, 6. 313-334. 10.18757/ejtir.2006.6.4.3455.
[7] Air Quality Expert Group, Air Quality and Climate Change: A UK Perspective, ISBN 0-85521-172-5, 2007.
[8] Shaaban S, and Seume J, Aerodynamic Performance of Small Turbocharger

Compressors. *Proceedings of the ASME Turbo Expo* 2007: Power for Land, Sea, and Air. Volume **3**: Turbo Expo 2007. Montreal, Canada. May 14–17 pp. 881-889, 2007. ASME.

## Acknowledgement

The research is part of the PhD program of the first author at the Politehnica University of Timisoara. The authors acknowledge also the Romanian Automotive Register for the substantial support during the measurements.

# The effect of using different Biodiesel fuels on jet development in a Diesel engine

**Andrei Laurentiu Niculae[1,*], Radu Chiriac[1,2], Alexandru Racovitza[1]**

[1]Faculty of Mechanical Engineering and Mechatronics, University Politehnica of Bucharest, Splaiul Independentei 313, 060042, Romania
[2]EA7341 CMGPCE of Conservatoire National des Arts et Metiers, Paris, France

*E-mail: andrei.niculae@upb.ro
ORCID: https://orcid.org/0000-0002-1174-289X

**Abstract**. The fuel properties and the injection rate-shape play an important role in the improvement of the combustion process of Diesel engines. In this work, the influences of using the forthcoming renewable biodiesel fuels on fuel jet development utilizing a computer simulation model created with the AVL Hydsim software were studied. Biodiesel fuels B20, B30 and B100 were considered and compared with the original pure Diesel fuel D100. The injection system behaviour under research was that one existing on a tractor engine equipped with Delphi DP200 pump and Delphi injectors. Two engine speeds of 1400 rpm and 2400 rpm were considered representative for the engine operation. For these speeds, the fuel jet characteristics as penetration, spray cone angle and Sauter mean diameter were analyzed. It can emphasize that in similar conditions of needle lift and injection rate-shape variation the usage of biodiesel fuels does not significantly alter the injection pressure and the Sauter mean diameter. However, the specific physical properties of biodiesel fuels affect substantially the spray penetration and its cone angle.

## 1. Introduction

The use of biodiesel has been presumed to become one of the most reliable ways to improve diesel engine performance and efficiency and to overcome the newest regulations regarding exhaust emissions levels. Due to appropriate fueling qualities, similar to diesel fuel, various mixtures of diesel-biofuel could easily operate in different diesel engine configurations with no significant modifications. However, the multitude of the involved parameters, starting with the fuel physical and chemical characteristics and continuing with those regarding the operation and the geometry of the injection strategy plays a key role in managing the processes of air-fuel mixture formation, combustion, heat release and emissions genesis.

A series of studies continue to assess the parametric relationship between the strategy input data and the followed efficiency and emissions parameters. One of these, as revealed by [1] is proposing to link by a linear relationship HC, $NO_x$ and smoke emissions concentrations to the volumetric proportion of the biofuel used. An AVL-5402, four-stroke, single-cylinder, normally aspirated, direct-injection diesel engine was tested in this study, fueled with neat diesel, B10, B20 and B30 and operated at 25%, 50% and 75% load and rating speeds of 1400 and 2200 rpm. An efficient linear connection between the volumetric percentage values of the fish-oil biodiesel blends, the fuel spray development and combustion, engine performance and emissions characteristics remains uncertain.

Another research direction refers to the injection spray characteristics as basic elements to influence the optimal injection timing and combustion delay (Fig.1 - Hawi et al. [2]). Their investigation highlights the effect of the injection pressure and ambient gas density on spray parameters when using neat diesel (D100) and methyl oleate biofuel (B100). The authors instrumented a rapid compression and extension machine, ensuring injection pressure values of 50, 100 and 150 MPa. It was concluded that the spray penetration increases with the increase of the injection pressure and decreases with the ambient gas density for both types of fuels. Also, improved atomization was indicated for all the tested fuels as a result of the liquid phase penetration decrease. Most significantly, biodiesel increases in vapour phase penetration, with a slight decrease of the spray angle while increasing the injection from 50 MPa to 150 MPa, indicating an improvement of the spray dispersion and leading to a better mixture quality. Nevertheless, the high values for viscosity and superficial tension in the case of biodiesel cause the reduction of the injection velocity, thus affecting the atomization characteristics of the fuel droplets [2-5]. This could be improved by controlling the ambient and the fuel temperatures [4,5], increasing the rate of the in-cylinder charge motion or mixing the biofuel with alcohol, whose lower viscosity and surface tension characteristics increase the interaction with the ambient gas [3,6].

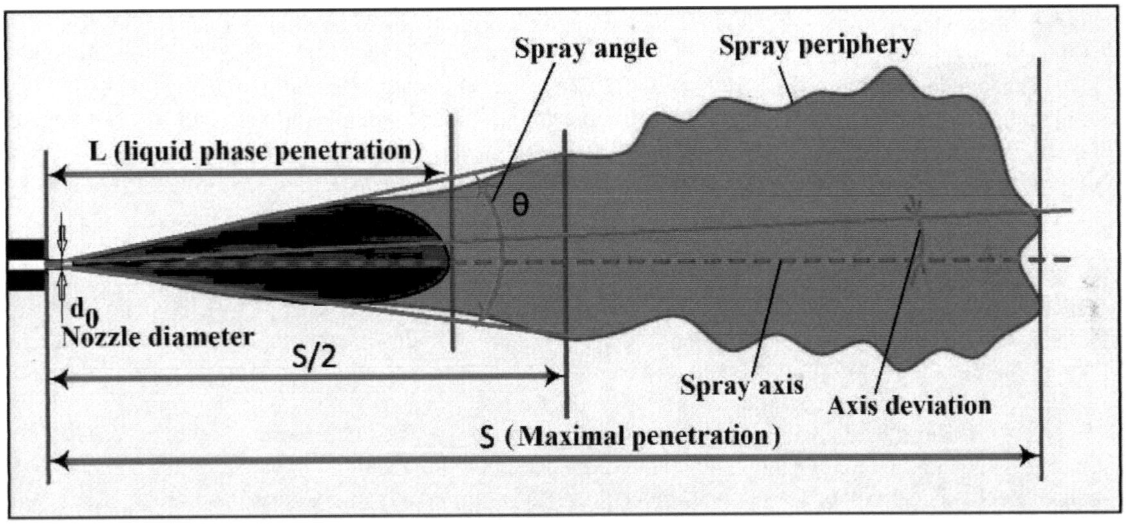


**Figure 1.** Spray geometry elements [2].

Similar tests and observations of the in-cylinder spray development had been conducted on extended ranges of engine loads and speeds. Lee et al. [7] carried out experiments on a Tata/Dicor 2.2L turbocharged diesel engine, equipped with a gas recirculation system and operating with a common-rail maximum injection pressure of 160 MPa, fueled with diesel and B40, B60 and B100 (v/v) biodiesel. Operating regimes were set at 100% load with 1500, 2000, 2500, 3000 and 3500 rpm and respectively at 2500 rpm with 0, 20, 40, 60, 80 and 100% load, while the injection pressure used was 50, 100 and 150 MPa and the in-cylinder pressure varied from 2 to 4 MPa. The findings suggested that at low injection pressure, diesel fuel showed a slightly higher injection rate compared to biodiesel blends. Oppositely, while fuel injection pressure increased, biodiesel revealed a higher injection rate, increasing even more with the blending ratio of biodiesel. Considering the entire operating range, longer spray penetration was noticed for biodiesel at higher injection pressure, with an extended trend by increasing the mixing ratio of biodiesel. At low ambient pressure, the penetration of B40 spray was shorter compared to that of diesel. For higher ambient pressure, B40 presented a similar evolution in terms of penetration and shape. At higher engine loads, the start of combustion (SOC) occurred at diesel before

B40. Biodiesel provided a longer combustion duration compared to neat diesel for all the tested regimes under 2500 rpm [7].

Together with the influence of the injection pressure, the number, the orientation, the opening and the shape of the nozzles are important parameters to complete the description of the injection characteristic. One particular study [8] analyzes the implications of the triangular cross-section of a single nozzle sprayer, equipping an experimental injector, operating under 50, 70 and 90 MPa pressure and injecting pure diesel (D100) and biofuel (B100) into a 99% pure Nitrogen ambience, with a pressure of 1, 2 and respectively 3 MPa. It was noticed that the use of biodiesel could inhibit the spray axis deviation. The higher probability of axis-switching phenomenon, the larger spray cone opening angle and spray width of diesel demonstrated a better system behaviour for diesel fuel at low injection pressure, and on the contrary, for biodiesel, at higher injection pressures [8]. Following a quite similar kind of research theme, L. Geng et al. [9] investigated the macroscopic and microscopic spray characteristics of biodiesel using a high-speed camera and a Malvern particle size analyzer (PSA). By decreasing the nozzle diameter, the spray angle for biodiesel decreased and its total penetration distance increased. Thus, the relative size range and dispersion boundary of the atomized biodiesel droplets increased significantly. Although decreasing the orifice diameter leads to the reduction of the mean diameter of the atomized droplets, the size distribution of the droplets appears not homogeneous [9].

The present work follows our previous simulation research undertaking on the comparison of performance, efficiency, and emissions of two tractor engine types, with four and three cylinders, with similar energetic configuration and same piston displacement, operated in classic diesel configuration [10]. The simulation applied two AVL-BOOST computation models, each of them using the AVL-MCC combustion equations system, calibrated on experimental data. Adding the scenario of both engines fueled by biodiesel B20, the main intention of the authors is to draw a parallel between the classic and the updated behaviour of the two engines, in terms of injection characteristics, simulated by the AVL HYDSIM tool.

## 2. Materials and Methods

Figure 2 consists of the symbolic model of the injection system provided by AVL Hydsim software. This program is an effective tool for the dynamic analysis of hydro-mechanical and hydraulic systems. Based on the theory of fluid dynamics and multi-body systems vibration, this software is mostly orientated to the simulation of the fuel injection [11].

The symbolic model of the injection system comprises as the main elements the injection pump Delphi DP200 and Delphi injectors which were provided on the 3 and 4-cylinder engines type 2302053 and respectively 2404056, maintained under production at 'Tractorul' factory from Brasov, Romania until 2007.

According to the manufacturer, the Diesel pump S – Delphi DP200 is a compact, distributor-type fuel injection pump designed primarily to meet the specialised requirements of naturally-aspirated and turbocharged diesel engines installed in light to medium duty, road-going vehicle, agricultural, tractor, industrial, and marine, applications [12]. Delphi manufactures DPS pumps in 2 or 4 pumping plungers configuration are suitable for 3, 4 and 6 cylinder engines configurations, of up to 1600 $cm^3$. The injection type can be either direct (DI) or indirect (IDI).

Figure 3 shows the design and the operating data of the radial piston distributor pump (RPD) Delphi DP200 and the nozzle orifice of Delphi injectors, as main elements of the injection system. Different values were introduced for configuring the RPD_Pump element, including the pressure in the cam chamber, roller radius, piston mass and diameter, as well as the cam's radial acceleration variation at 1000 rpm. To configure the Noozle_Orifice element, geometric dimensions were introduced regarding the flow through the 5 spray holes, these having a diameter of 0.24 mm and length of 1 mm.

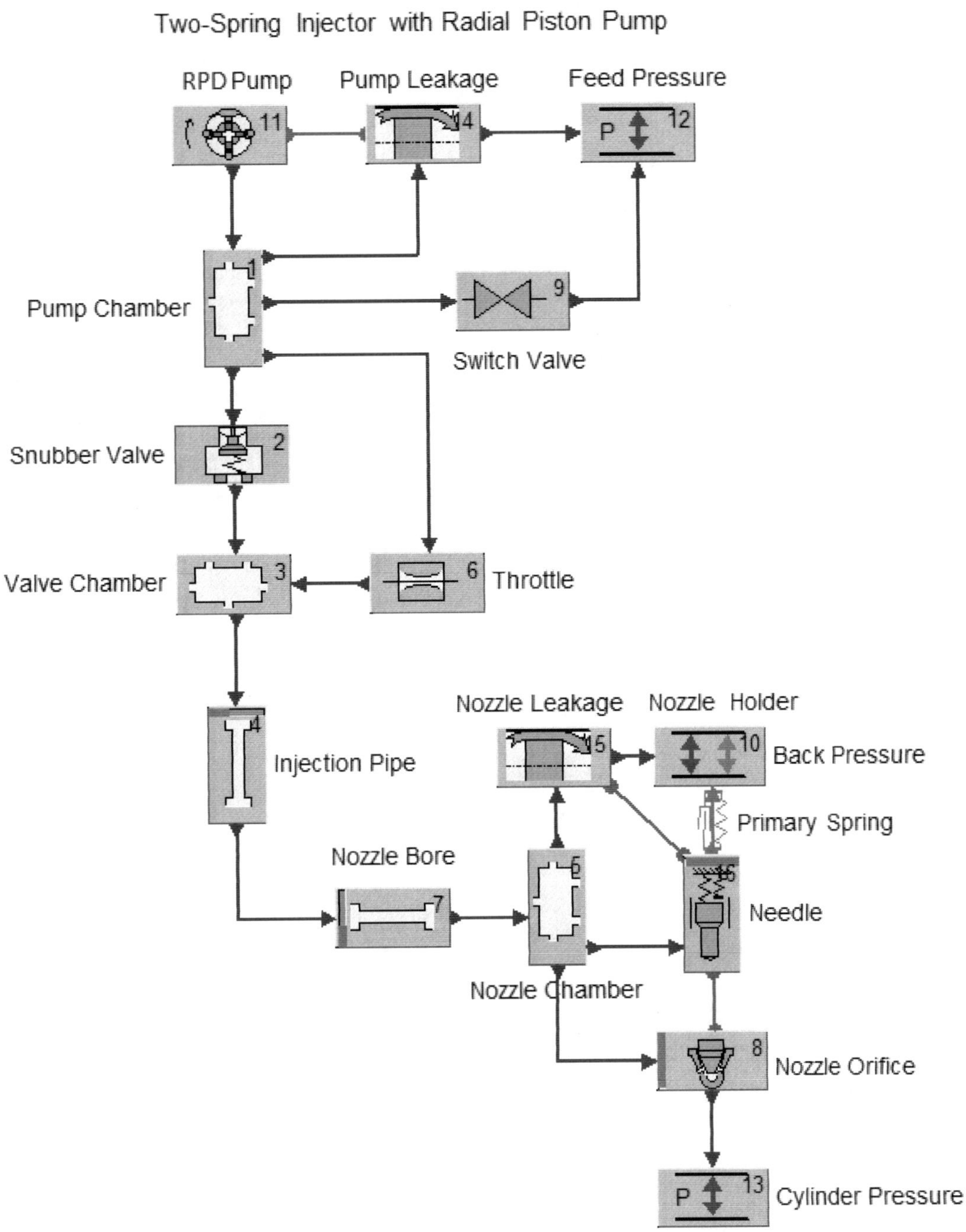


**Figure 2.** The symbolic generated model of the injection system using AVL Hydsim software.

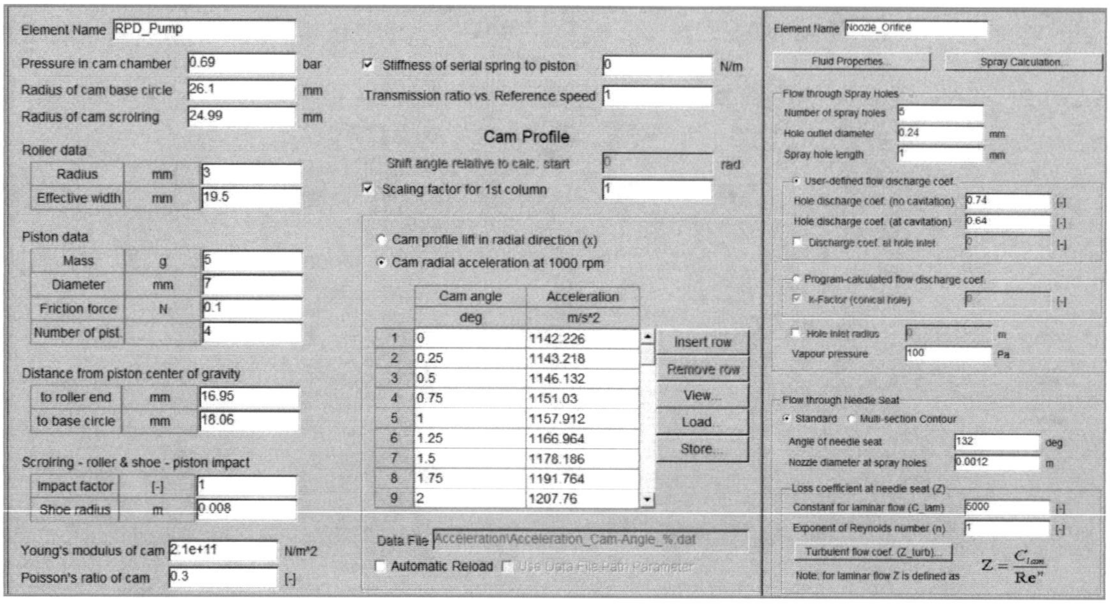


**Figure 3.** The design and the operating data of the Delphi DP200 injection pump and injectors.

Significant results in the study of injection characteristics could be obtained by using this program. The variations of the fuel properties related to the pressure are presented in Figure 4 for pure diesel fuel D100, respectively pure biodiesel B100 and in Figure 5 for B20 and B30 biodiesel blends. These are necessary to establish the fuel spray shape that influences the formation of the air-fuel mixture and finally the combustion of the premixed charge.

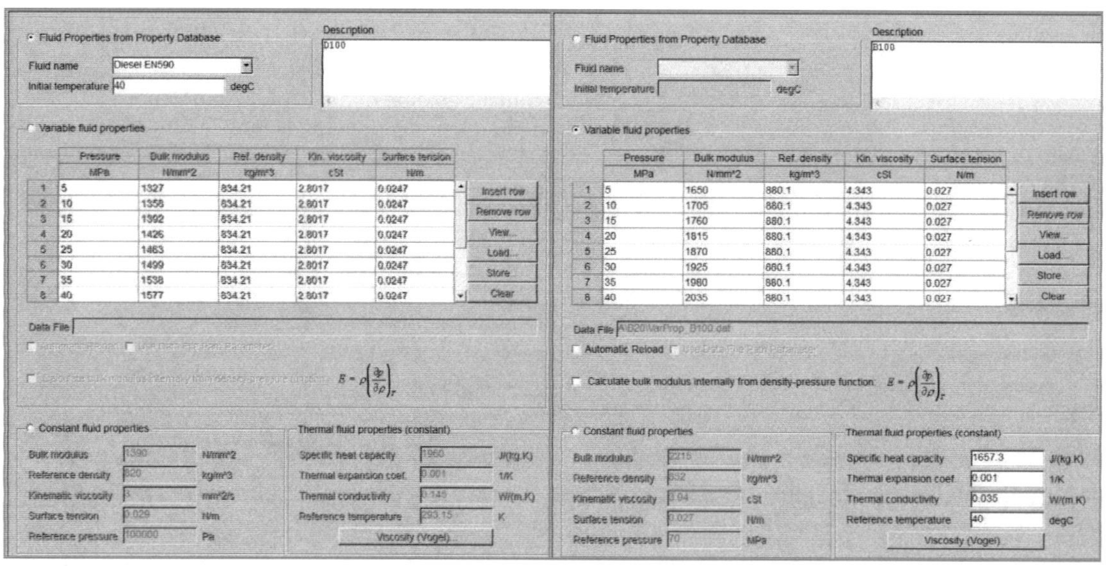


**Figure 4.** Properties of pure diesel (D100) and biodiesel (B100) fuels.

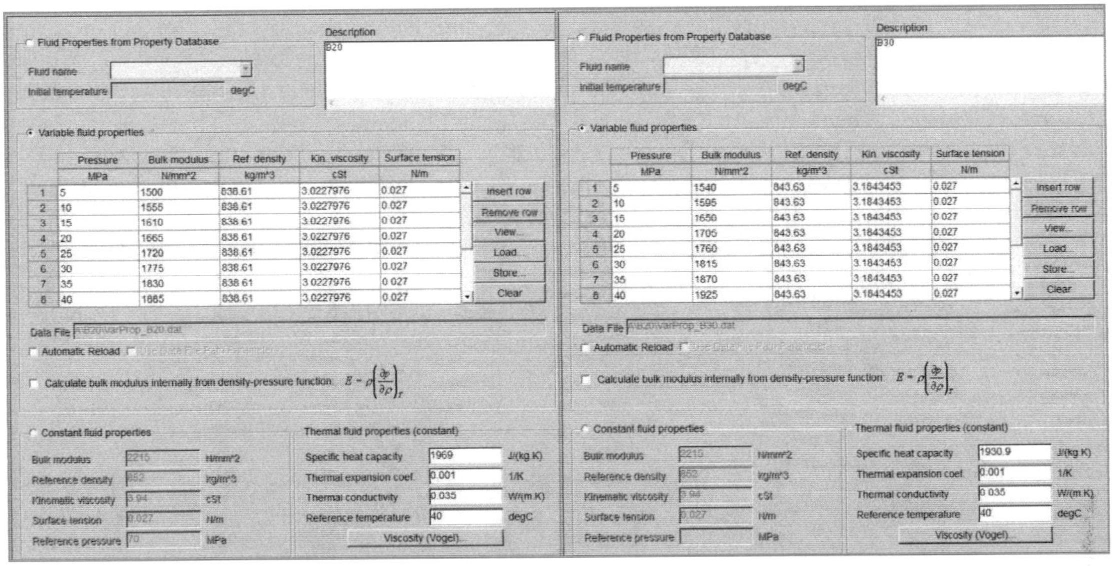


**Figure 5.** Properties of B20 and B30 biodiesel blends.

Sensitive parameters of the injection process were evaluated taking into account two operating speeds, 1400 rpm and 2400 rpm. These are in the range of operating speeds of the two engines. For the 4-cylinder engine type 2404056, the 1400 rpm speed corresponds to the maximum torque, while the other one 2400 rpm speed generates the maximum power.

Figure 6 displays the variation of needle lift and injection rate versus the crank angle. Considering both engine speeds, it was noticed that the maximum value of the needle's lift remains constant regardless of the fuel used. There is a small 0.5 deg advance of the needle's lift when using B100 at 2400 rpm. This advance decreases as the percentage of biodiesel in fuel goes down. Also, this advance is lower at 1400 rpm. Regarding the injection rate, the use of biodiesel affects both the maximum value and the shape of the variation. When using biodiesel B20, B30 or B100, there is an increase in peak value relative to pure diesel D100, by 5.7% when considering 1400 rpm speed and 5.5% for 2400 rpm speed. The start of injection is also advanced when using biofuels with 0.2 deg at 1400 rpm and 0.4 deg at 2400 rpm.

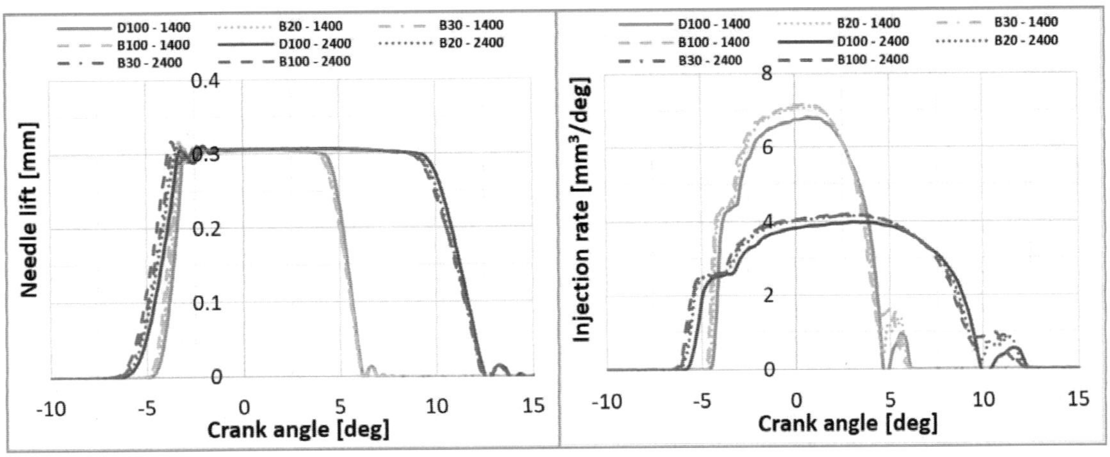


**Figure 6.** The injection rate and needle's lift variation vs. the crank angle.

Figure 7 shows the variation of the injection pressure and Sauter mean diameter versus the crank angle. The maximum pressure values are recorded for both speeds when using B100 fuel, with an increase of 15.2% relative to D100. There is also the same tendency of the injection pressure variations to move in advance when using biodiesel, the higher being the biodiesel ratio the greater being the advance of the pressure characteristic. During the injection process, the fuel does not sensibly influence the Sauter mean diameter values. However, the maximum values of 0.186 mm are recorded when using B20 for both engine speeds. The injection process can be divided as follows:

- the initial phase (from the start of injection to -3 deg for 1400 rpm, respectively to -4 deg for 2400 rpm);
- the main phase (between -3 and +4 deg for 1400 rpm, respectively -4 and +8 deg for 2400 rpm);
- the final phase (the rest of the injection period).

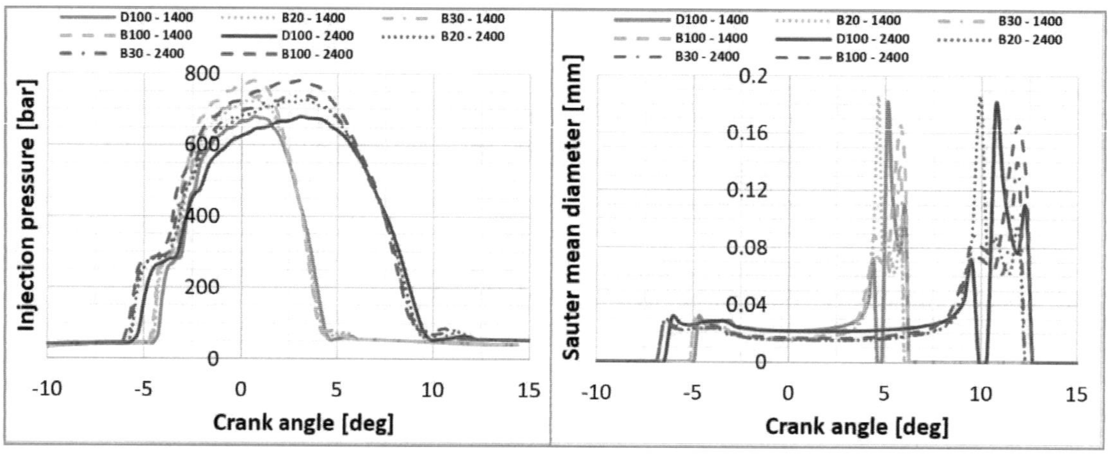


**Figure 7.** The spray angle and Sauter diameter vs. the crank angle.

Considering the initial injection phase, the average value of the Sauter mean diameter was 0.0252 mm at 1400 rpm and 0.0256 mm at 2400 rpm, while for the main phase it was 0.0211 mm for 1400 rpm and 0.0195 mm for 2400 rpm and for the final phase it was 0.0869 mm at 1400 rpm and 0.0724 mm at 2400 rpm. Also, the same average value of 0.0330 mm was calculated for both speeds considering the entire injection period. It can also be noticed that there are no substantial differences between the fuels considered and that the increase of biofuel fraction in biodiesel fuels does not significantly change the average value of the Sauter mean diameter.

Figure 8 highlights the variations of the spray penetration and angle versus the angular position of the crankshaft. Considering the engine operating at both speeds, these variations are similar for both characteristics of the fuel spray. However, the shift from the use of D100 to biodiesel B20, B30 and B100 is associated with a significant increase in the fuel spray penetration. The maximum value increases from 0.0158 m for D100 to 0.0271 m for B20. Also, the spray angle increases significantly with the increase of biodiesel fraction in fuel. Thus, its maximum value increases from 8.8 deg for D100 to 13.9 deg for B100, 17.3 deg for B30 and 17.8 deg for B20, for both speeds. The variations of the spray angle are similar for both speeds.

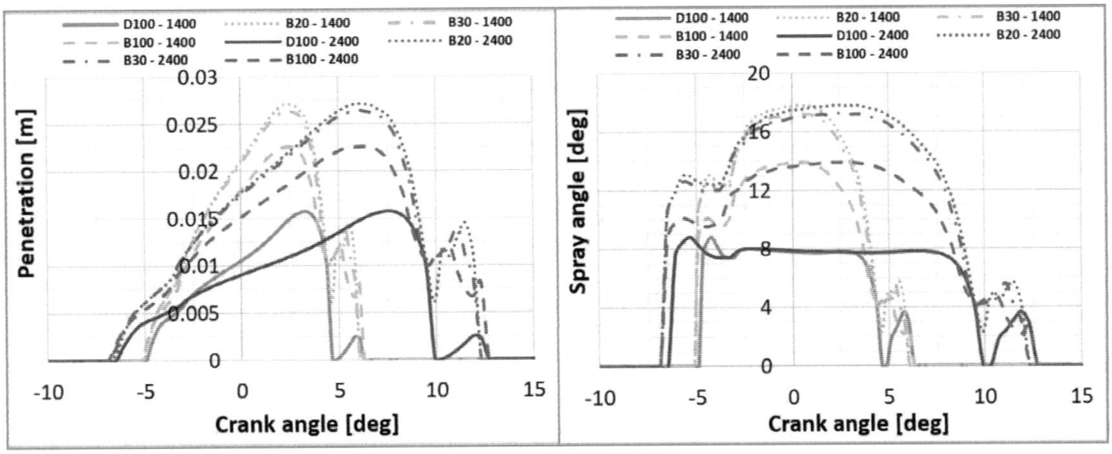


**Figure 8.** The penetration and spray angle vs. the crankshaft angle.

## 3. Conclusion

The results obtained in this study could be summarized as the following conclusions:

- the injection system is a very complex hydro-mechanical system for which the fuel physicochemical properties have a certain influence on its normal functioning;
- the mixture formation and the combustion qualities in Diesel engines are strongly correlated with the fuel jet characteristics;
- the changing of the original pure diesel fuel D100 with biodiesel fuels B# containing different fractions of biofuel in diesel fuel can modify the geometry of the fuel jet, particularly the spray penetration and its cone angle;
- the modification of these spray parameters seems to be not in a direct linear correlation with the biofuel fraction in biodiesel, the highest values being obtained for B20 at both engine speeds considered;
- these results should be cautiously regarded because they have to be verified on other models and obviously on experimental databases.

**Acknowledgements**

The authors of this paper acknowledge the AVL-AST team from AVL List GmbH for the special support offered under the University Partnership Program on performing the simulation activity.

## References

[1] Nguyen T, Pham M H and Anh T Le 2020 Spray, combustion, performance and emission characteristics of a common rail diesel engine fueled by fish-oil biodiesel blends *Fuel* **269** 117108

[2] Hawi M, Kosaka H, Sato S, Nagasawa T, Elwardany A and Ahmed M 2019 Effect of injection pressure and ambient density on spray characteristics of diesel and biodiesel surrogate fuels *Fuel* **254** 115674

[3] Algayyim S J M and Wandel A P 2021 Macroscopic and microscopic characteristics of biofuel spray (biodiesel and alcohols) in CI engines: A review *Fuel* **292** 120303

[4] Anis S and Budiandono G N 2019 Investigation of the effects of preheating temperature of biodiesel-diesel fuel blends on spray characteristics and injection pump performances *Renew. Energy* **140** 274–80

[5] Suraj C K, Sudarshan G, Anand K and Sundararajan T 2021 Effects of autooxidation on the fuel spray characteristics of Karanja biodiesel *Biomass and Bioenergy* **149** 106084

[6] Geng L, Bi L, Li Q, Chen H and Xie Y 2021 Experimental study on spray characteristics, combustion stability, and emission performance of a CRDI diesel engine operated with biodiesel–ethanol blends *Energy Reports* **7** 904–15

[7] Lee S, Lee C S, Park S, Gupta J G, Maurya R K and Agarwal A K 2017 Spray characteristics, engine performance and emissions analysis for Karanja biodiesel and its blends *Energy* **119** 138–51

[8] Yu S, Yin B, Deng W, Jia H, Ye Z, Xu B and Xu H 2018 Experimental study on the diesel and biodiesel spray characteristics emerging from equilateral triangular orifice under real diesel engine operation conditions *Fuel* **224** 357–65

[9] Geng L, Wang Y, Wang Y and Li H 2020 Effect of the injection pressure and orifice diameter on the spray characteristics of biodiesel *J. Traffic Transp. Eng. (English Ed.)* **7** 331–9

[10] Niculae A L, Chiriac R, Racovitza A and Ungureanu V 2021 Effect of injection system characteristics on performance, efficiency and emissions of agricultural tractor Diesel engines *paper submitted to "Automotive and Integrated Transport Systems – AITS 2021" conference, Chisinau, Rep.of Moldova*

[11] AVL Hydsim Theory 2019

[12] Delphi – Principles of Operation – DPS Fuel Injection 2009

# Study concerning the thermal energy balance in an internal combustion engine

**RM POPA, AT BORBOREAN[0000-0001-6818-9549], V STOICA[0000-002-7945-7817], and I IONEL[0000-0002-2999-1210]**

Politehnica University of Timisoara, Faculty of Mechanical Engineering, Timisoara, Piața Victoriei 2, 30006, Romania

E-mail : ioana.ionel@upt.ro

**Abstract.** In this scientific report the steps for achieving the energy balance on a test engine Daewoo 1.6 liter in 4 cylinders, with 4 valves per cylinder, its power being77 kW, are presented. The engine was connected to a hydraulic brake during the tests. The energy balance was established, based on measurements applied on the stand test, focusing to determine the lost heat and to conceive a Sankey diagram, accordingly. Two sets of measurements were performed and based on them the heat losses were determined and two Sankey diagrams were depicted. The lost heat through the cooling fluid is determined by measuring the cooling water flow rate and its inlet and outlet temperature. The lost heat through the exhaust gases is determined by the difference between the enthalpy of the exhaust gases and the enthalpy of fresh air. The lost heat due to incomplete chemical combustion is determined by analyzing the composition of the exhaust gases. The measurements are centralized on a graphic user interface, supported by a NI Compact RIO platform.

## 1. Introduction and scope of the research

The purpose of this paper is to determine a calculation model and steps to achieve the thermal balance of an engine. It presents how the energy balance was achieved on a 1.6-liter Daewoo test engine, in 4 cylinders with 4 valves per cylinder, its power being 77 kW. The engine was connected to a hydraulic brake. Thermal energy balance and energy management are very important in the relationship with real driving emissions and of course improving the fuel economy. At internal combustion engines, heat loss, due to heat exchange of the combustion gases with the cylinder wall and the coolant liquid, influences the thermodynamics of the engine cycle [1]. Heat loss identification is a key part of the energy balance concept, beside gas temperature and pressure influences upon the engine performance and emissions [2].

Another goal of this paper was to pursue and discover energy recovery solutions. The thermal balance is useful in the calculation for the design of the cooling system and the calculation of the exhaust gas recovery system [3].

One of the best solutions of the energy recovery is the turbo-compressor group, which takes over the gases and uses them; more precisely it uses the energy of the exhaust gases to introduce fresh air into the engine with a higher flow than the engine would receive normally [4].

The strongest arguments for energy recovery are closely related to energy demand. Placing waste in landfills clearly leads to a waste of energy that could be usefully used. In addition, land and nature are used inefficiently [3]. The arguments concerning thermal energy recovering solutions are based in generally on emissions and toxic substances, especially carbon dioxides, which are harmful to the planet and man by default. However, an estimated calculation showed that incineration is equal in value, perhaps even lower compared to material recovery [3].

## 2. Presentation of the experimental stand

The energy balance was established based on measurements applied on the test rig in the engine laboratory of the Faculty of Mechanical Engineering, within the Politehnica University of Timisoara (Figure 1). Technical specification of the engine used on these experiments is following: 4 cylinder in line DOHC Daewoo 1598 cm$^3$, bore × stroke 79 × 81.5 [mm], compression ratio 9.5. With the help of this test rig it was possible to establish the engine thermal balance. The tests could be performed depending on the operating modes. Figure 1 presents the entire experimental rig with the internal combustion engine mounted. The rig contains some sensors like knock sensor, oil pressure sensor, two tanks, one of water and one of fuel, and a control panel from where the entire experimental stand was controlled. The stand has and a NI computer who helped to acquire all the entire measurements and to collect them.

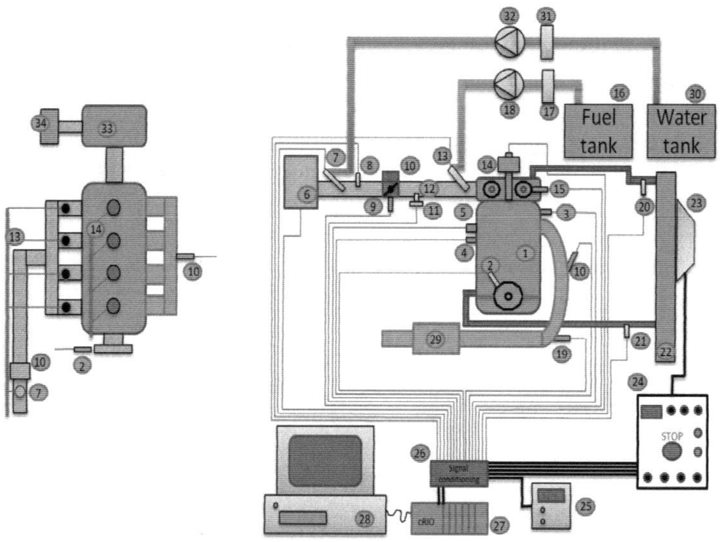


**Figure 1. Experimental test rig** [5].

1. Internal combustion engine; 2. Crankshaft position sensor; 3. Temperature cooling sensor; 4. Oil pressure sensor; 5. Shock sensor; 6. Mass air flow sensor; 7. Water injector; 8. Intake air temperature sensor; 9. The position sensor of the regulator; 10. Valve position; 11. Inlet air pressure sensor; 12. Intake gallery; 13. Fuel injector; 14. Ignition coil; 15. Camshaft position sensor; 16. Fuel tank; 17. Fuel filter; 18. Fuel pump; 19. Exhaust temperature sensor; 20. Intake cooling radiator temperature sensor; 21. Exhaust cooling radiator temperature sensor; 22. Heat exchanger; 23. The fan; 24. Control panel; 25. Power supply; 26. Signal change box; 28. Computer; 29. Evacuation gallery; 30. Water tank; 27. Crio national instruments; 31. Water filter; 32. Water pump; 33. Water brake; 34. Scale.

The test rig is also connected to a PC unit, where there is also a graphical interface with who can monitoring and control the engine in real time. The graphical interface was designed in National Instruments NI LabVIEW. The interface indicates obtained results, those being further used in the calculations. These refer to the different inlet or outlet temperatures of the engine air and water temperatures, speed (RPM), pressures in engine, the amount of fuel, fuel consumption, specific fuel consumption, power engine, the force at the balance, the position of the damper, spark advance and

Dwell angle. In addition, thanks to a gas analyzer that is running on the test rig, the analysis of carbon monoxide and hydrocarbons, oxygen and nitrogen oxides could be performed.

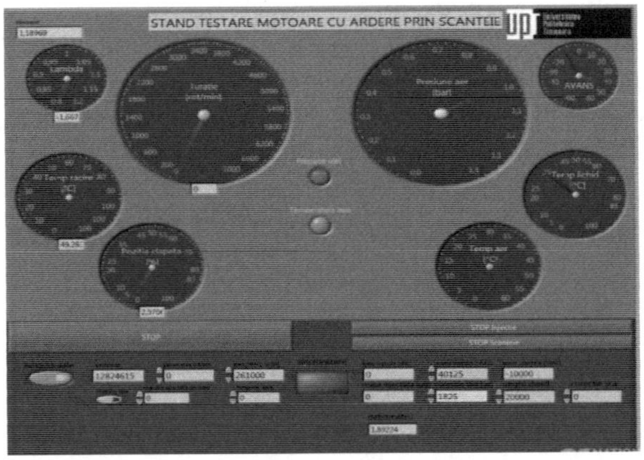


**Figure 2. Graphical interface of the experimental stand** [6].

In Figure 2 is presented the graphical interface of the experimental stand which includes indicators for the excess air coefficient, the engine rotation, the air pressure and temperature and the coolant temperature. Figure 3 indicates a general view of the experimental test rig, as used.

**Figure 3. Experimental test rig.**

## 3. Results and comments concerning the experimental achieved energy balance.

Usually the thermal balance is established for a specific characteristic of the engine, this being in most cases a characteristic of speed, load or regulator. The measurements wereinitiated only when the engine reached the desired speed (3000 rpm). One mention that was very important to respect was the next condition: when measuring the water flow with the water meter, the temperature of the cooling water must not exceed 90 °C. More exactly, the temperature of the cooling water has been kept around 75°. Another mention that is important to specific is the one about the engine load which is partial load, throttle position was at 15%, and the RPM was constant.

In the case of the established operating regime, an hourly fuel consumption had to be determined, the hourly air flow, the cooling water flow was also measured (the procedure being based by a timing of the interval in which the water meter indicators covered a certain number of divisions), simultaneously flue gases were analyzed. At certain time intervals, being either 30 or 60 seconds, the following were read the engine rotation, braking force, the next following temperatures like the intake air in the engine, exhaust gases and the air pressure when the air enter in the engine.

The engine operating conditions for the two sets of measurements are pretty much the same with the mention that in the set with $q_{cool}$ = 20.24% the fan on the heat exchanger (radiator) is on, while in the second set of measurements it is off.

Relation (1) presents the formula for available heat flow which contain the amount of all the lost heat flows

$$\dot{Q}=\dot{Q}_e+\dot{Q}_{cool}+\dot{Q}_g+\dot{Q}_{in}+\dot{Q}_{res}\ [2] \tag{1}$$

where: $\dot{Q}$-is the heat flow of the applicators of the calorific value of the consumed fuel (available heat);

$\dot{Q}_e$-- heat flow transformed into effective mechanical work;

$\dot{Q}_{cool}$ - heat flow dissipated by the cooling fluid;

$\dot{Q}_g$ - heat flow evacuated with flue gases;

$\dot{Q}_{in}$ - heat flow evacuated with flue gases due to incomplete combustion;

$\dot{Q}_{res}$ - the residual term that includes losses that are not part of the other terms of the heat flow balance [2].

*3.1. Determination of energy losses through the cooling system*

The energy loss achieved by the cooling system ($\dot{Q}_{cool}$) is given by the following formula (2):

$$q_{cool} = \frac{\dot{Q}cool}{\dot{Q}}\,100 \tag{2}$$

where: $\dot{Q}_{cool} = C_p \times \dot{M}_{cool} \times (t_2 - t_1)$;

$C_p[\frac{kJ\ /\ kg}{grd}]$ – specific heat at a constant coolant pressure;

$\dot{M}_{cool}[kg/h]$ –coolant flow;

$t_1$ , $t_2$ –coolant temperatures at the inlet and outlet of the engine [2].

*3.2. Determination of energy losses through exhaust gases*

$\dot{Q}_g$ is the heat flow dissipated with flue gases and is calculated with relation (3).

$$\dot{Q}_g = \dot{M}_g \times i_g - (\ \dot{M}_a \times i_a + C \times i_c) \quad [\text{kJ/h}] \tag{3}$$

where: $\dot{M}_g = \dot{M}_a + C$ *[kg/h]* – exhaust gas flow;

$\dot{M}_a$ *[ kg/h ]* -airflow allowed;

$C$ *[kg/h ]* - hourly fuel consumption;

$i_g$ *[kJ/kg]* - specific enthalpy of gases;

$i_a$ *[kJ/kg]* - idem for air;

$i_f$ *[kJ/kg]* - idem for fuel [2].

*3.3. Estimation of energy losses by lubrication and radiation system*

The lubrication system has the role of increasing the service life of the engine due to the reduction of friction. The higher the friction, the more intense is the wear process. Friction decreases when the surfaces of the friction parts will be separated by an intermediate layer of liquid with specific lubrication properties. Friction losses depend on lubrication conditions. The use of too viscous oil increase the average pressure, in the same way acting less viscous oil which due to the film has a lower load-bearing capacity, is interrupted and allows the appearance of a dry and semi-liquid friction. Heat lost through radiation and convection is due to the outer surfaces of the equipment [2], [5].

*3.4. Heat available*

The available heat flow is calculated with relations (4a, 4 b) [6, 7, 8].

$$\text{If } \alpha \geq 1: \dot{Q} = C \times H_i \ [kJ/h] \tag{4a}$$

$$\text{If } \alpha < 1; \dot{Q} = C\,(H_i - \Delta H_i)\ [kJ/h] \tag{4b}$$

where: ΔHi - heat lost through incomplete combustion.

In the case of petrol: Hi ⁼ 43.58 × 103 kJ / kg[2] ;

Air fuel ratio is: λ = ~ 1;

BMEP = ~ 3.1 bar .

*3.5. Heat transformed into effective mechanical work*

*Pe [kW]* $\dot{Q}_e = 3.6 \times 103 \times Pe\ [kJ/h]$;

*Pe [CP]* $\dot{Q}_e = 632.3\ [kcal/h]$[2].

*3.6. Heat dissipated by coolant*

$$\dot{Q}_{cool} = C_p \times M_{cool} \times (t_2 - t_1) \tag{5}$$

where: Cp $[\frac{kJ/kg}{grd}]$is the specific heat at a constant pressure of the cooling fluid;

$M_{cool}$ [kg / h] - coolant flow rate;

$t_1$, $t_2$[° C] - coolant temperatures at the inlet and outlet of the engine, respectively.
If the exhaust manifold is cooled, the heat taken from the manifold will also be taken into account [2].

*3.7. Evacuated heat with exhaust gases*

The evaluation of the evacuated heat with exhaust gases is based on formula 6.

$$\dot{Q}_g = C_{pgm} \times (\dot{M}_a + C_c) \times (t_g - t_e)\ [kJ/h] \tag{6}$$

where: $C_{pgm}[\frac{kJ/kg}{grd}]$ is the average specific heat of the flue gases;

$C_{pgm} \approx 1.172\ kJ/(kg \times grd)$- for petrol engines;

$t_g$ [° C] - temperature of the exhaust gases;

$t_e$ [° C] - the air temperature at the engine inlet[2].

*3.8. Heat lost through incomplete combustion*

The heat flow, lost through incomplete combustion, is estimated with the formula (7)[6,7].

$$\dot{Q}_{in} = 22.4 \times n_g \times C \times H_{ig} + \varphi_c \times c \times C \times H_{ic} \tag{7}$$

where: $n_g = k_{mol} / (kg \times comb)$ is the number of $k_{mol}$ of gas per kg of fuel;

$H_{ig}$ $[kJ / Nm^3]$ - lower calorific value of gases;

$\varphi_c$ - the fraction of unburned carbon;

$c$ - $[kgC / kg \times comb]$ - the amount of carbon in a kg of fuel;

$H_{ic}$ $[kJ / kg]$ - lower calorific value of carbon[2].

*3.9. Residual heat*

The residual heat flow is calculated with relation (8).

$$\dot{Q}_{res} = \dot{Q} - (\dot{Q}_e + \dot{Q}_{cool} + \dot{Q}_g + \dot{Q}_{in}) \quad [kJ/h] \tag{8}$$

where: $\dot{Q}_e$ - heat flow transformed into effective mechanical work;

$\dot{Q}_{cool}$ - heat flow dissipated by the cooling fluid;

$\dot{Q}_g$ - heat flow evacuated with flue gases;

$\dot{Q}_{in}$ - heat flow evacuated with flue gases due to incomplete combustion;

$\dot{Q}_{res}$ - the residual term that includes losses that are not part of the other terms of the heat flow balance [2].

Tables 1 and 2 show the two sets of measurements results obtained on the experimental stand.

**Table 1.** The results for the first set of measurements taken on the experimental stand.

| $\dot{Q}$ [kJ/h] | $\dot{Q}_{ef}$[kJ/h] | $\dot{Q}_{cool}$ [kJ/h] | $\dot{Q}_{gas}$ [kJ/h] | $\dot{Q}_{res}$ [kJ/h] |
|---|---|---|---|---|
| 250680.876 | 44480.62019 | 34974.74915 | 76843.13983 | 94382.36683 |
| 250680.876 | 44486.59526 | 40598.15836 | 77304.07459 | 88292.04779 |
| 251229.984 | 44621.03867 | 58193.11501 | 77536.0031 | 70879.82722 |
| 251229.984 | 44580.6204 | 65258.95911 | 77740.78149 | 63649.62314 |
| 252563.532 | 44639.70069 | 55319.88446 | 77892.07391 | 74711.87294 |

IOP Conf. Series: Earth and Environmental Science **960** (2022) 012012 doi:10.1088/1755-1315/960/1/012012

**Table 2.** The second set of measurements that has been taken on the experimental stand.

| $\dot{Q}$ [kJ/h] | $\dot{Q}_{ef}$[kJ/h] | $\dot{Q}_{cool}$ [kJ/h] | $\dot{Q}_{gas}$ [kJ/h] | $\dot{Q}_{res}$ [kJ/h] |
|---|---|---|---|---|
| 250628.58 | 44648.30621 | 50225.7691 | 78060.38621 | 77694.1184 |
| 250628.58 | 44616.8812 | 38215.25391 | 78030.50841 | 89765.9364 |
| 252798.864 | 44653.20103 | 35517.558 | 78022.99675 | 94605.1082 |
| 252798.864 | 44686.31137 | 32398.9298 | 77972.36977 | 97741.2530 |
| 252641.976 | 44582.86395 | 30741.44682 | 77956.28053 | 99361.384 |
| 252641.976 | 44674.9308 | 32575.60094 | 77948.58815 | 97442.8561 |
| 246627.936 | 44568.9852 | 30555.05539 | 77855.61359 | 93648.3670 |

The tables 3 and 4 present the averaged percentage representation of the two sets of measurements.

**Table 3**. The results for the first set of measurements.

| | |
|---|---|
| q | 100% |
| $q_{ef}$ | 17.73% |
| $q_{cool}$ | 20.24% |
| $q_g$ | 30.83% |
| $q_{res}$ | 31.19% |

**Table 4.** The results for the second set of measurements.

| | |
|---|---|
| q | 100% |
| $q_{ef}$ | 17.76% |
| $q_{cool}$ | 14.23% |
| $q_g$ | 31.04% |
| $q_{res}$ | 36.97% |

The engine operating conditions for the two sets of measurements are pretty much the same with the mention that in the set with $q_{cool}$ = 20.24% the fan on the heat exchanger (radiator) is on, while in the second set of measurements it is off.

The diagrams in Figures 4 and 5 are accomplished after the two sets of measurements:

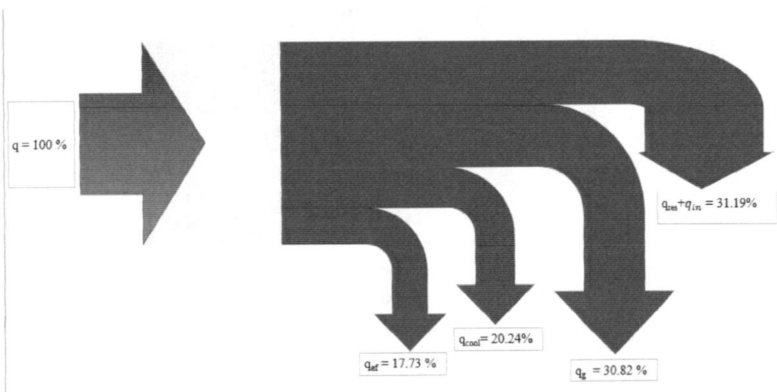


**Figure 4.** Sankey diagram for the first set of measurements.

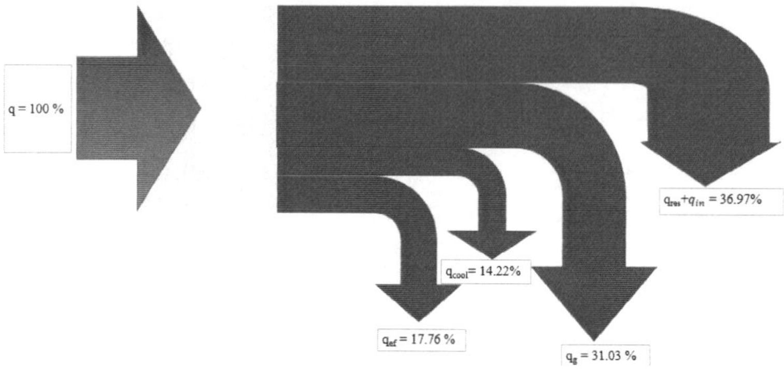


**Figure 5.** Sankey diagram for the second set of measurements

## 4. Comments regarding the results

The purpose of this paper was to create an energy balance for the Daewoo 1.6l engine with a power of 77 kW and to create a Sankey diagram to express the available heat, the heat transformed into mechanical work, the heat evacuated by the cooling system, the heat evacuated by the exhaust gases, heat lost through incomplete combustion and residual term heat.

The results of the measurements, expressed as a percentage from the total flow, are as follow:

- $q_{ef}$ = 17.73% for the first set of measurements, $q_{ef}$= 17.76% for the second set of measurements;
- $q_{cool}$ = 20.24% for the first set of measurements, $q_{cool}$ = 14.23% for the second set of measurements;
- $q_g$ = 30.83% for the first measurement set, $q_g$ = 31.04% for the second set of measurements;
- $q_{res}$ = 31.19% for the first set of measurements, $q_{res}$ = 36.97% for the second set.

The energy balance had the role of analyzing the heat losses that existed during the test. With the help of this energy balance was received some important clarifications regarding the energy losses by locating and clearly establishing the true causes of the losses. Based on these measurements is possible to evaluate the possibilities to recover a part of the dissipated heat, as long as we already know the source, the quantity and the quality of the dissipated heat.

## Conclusion

This paper performs a thermal energy balance on an internal combustion engine. The results obtained are similar to those in the literature. Sankey diagrams made with the help of the two sets of measurements show the situation and the representation of lost heat flow.It can be seen that the largest heat losses were residual heat and heat lost due to incomplete combustion for both sets of measurements.

## Bibliography

[1] Fonseca L O R N, Internal combustion engine heat transfer and wall Temperature,*Arch Computat Methods Eng* , 2020.

[2] Berindean V, Internal combustion engine, Timisoara: Politehnica Institute, 1981, p. 63 (Original in Romanian).

[3] Boding P A H, Renewable energy from waste through gasification or incineration for heat, power or fuel production, vol. 1, 2003.

[4] Cancel L, www.e-automobile.ro, [Online]. Available: http://www.e-automobile.ro/categorie-

motor/20-general/33-motor-turbo.html. [Accessed 08 2021].

[5] Borborean AT, Test bench for studying the effects of water injection inside an internal combustion engine, 2015, Licence thesis, Universitatea Politehnica Timişoara, (Original in Romanian).

[6] Stoical V, Minimal algorithm for running an internal combustion engine, *IOP Conference Series Materials Science and Engineering* , 2018.

[7] Grunwald, Theory, calculation and construction of engines for road vehicles, Bucharest: Didactic and Pedagogical Publishing House, 1980 (Original in Romanian).

[8] Berry C W . G, Heat transfer in internal combustion engines,*The American society of mechanical Enginees (ASME),* 1985.

**Acknowledgement**

The paper is accomplished in the frame of the PhD program of the first author.

# Turbocharger surge behavior for sudden valve closing downstream the compressor and effect of actuating variable nozzle turbine

**A Danlos[1*], P Podevin[1], M Deligant[2], A Clenci[3], P Punov[4] and S Guilain[5]**
[1]Le Cnam, Arts et Metiers Institute of Technology, LIFSE, HESAM Université, F-75003 Paris, France
[2]Arts et Metiers Institute of Technology, Le Cnam, LIFSE, HESAM Université, F-75013 Paris, France
[3]Department Automobiles and Transport, University of Pitesti, 110040 Pitesti, Romania
[4]Department of Combustion Engines, Automobile Engineering and Transport Université technique de Sofia, Sofia, Bulgaria
[5]Renault SA, Powertrain Advanced Engineering, France

*amelie.danlos@lecnam.net ORCID ID : 0000-0001-7598-7434

**Abstract**. Surge is an unstable phenomenon appearing when a valve closing reduces the compressor flow rate. This phenomenon is avoided for automotive turbochargers by defining a surge margin during powertrain system design. This surge margin established with measurements in steady state testing regime limits the maximal engine torque at low levels of output. An active control of the compressor could reduce the surge margin and facilitate a transient compressor operation for a short time in surge zone. In this paper, an experimental study of the transient operation of a turbocharger compressor entering the surge zone is performed. Control of the turbocharger speed is sought to avoid unsteady operation using the variable geometry turbine (VGT) nozzle actuator. From a given stable operating point, surge is induced by reducing the opening of a valve located downstream of the compressor air circuit. The effect of reducing the speed of rotation by controlling the VGT valve is investigated, as this should lead to more stable compressor operation. The rotation speed of the turbocharger is controlled to avoid an unstable operating point using servo-actuator of variable geometry turbine. From a stable operating point, the surge appearance is caused by closing a butterfly valve downstream the air circuit of the compressor. The effect on the compressor rotation speed when the opening of variable geometry turbocharger valve is modified is studied. Measurements have been conducted for different control profiles of the VGT valve placed downstream the compressor. This article presents the means used to carry out these tests as well as the results of the measurements of the instantaneous signals of pressure, temperature, flow rate and rotation speed, allowing the analysis of the surge phenomenon.

## 1. Introduction

Automotive manufacturers use chargers in powertrain systems for reducing $CO_2$ emissions in internal combustion engines or for increasing performances of proton exchange membrane (PEM) fuel cells as new power of resource [1]. The operating range of compressors in this type of applications is limited by rotating stall leading to surge. Surge is an instability which appears at low flow rates and high load. The appearance of this phenomenon may cause damages to the turbocharger but also mechanical failures on internal combustion engine if relevant procedures are not applied to avoid it. Surge has been studied for last three decades [2, 3] to design systems to avoid its occurrence [4, 5]. Many parameters affect the surge instability and especially geometry and operating conditions of the compressor but also downstream and upstream geometry circuit [6-10].

Surge limit is determined by stationary experiments on test bench. But on the road, inside the vehicle, this limit can vary because of high altitude [11] or acceleration and deceleration [12]. It is then important to define a criterion to detect the surge limit in use, to adjust and control powertrain for low flow rates. Different methods are suggested in literature: temporal analysis of pressure ratio or mass flow rate, frequency analysis, inlet temperature survey [13-16]. As transient behaviour of compressor is crucial to analyse to prevent surge to appear [17], the detection of surge limit with standard criteria needs adjustments to validate the efficiency of methods [18].

In this paper, an experimental method to control surge in transient operation of turbocharger is proposed, managing variable geometry turbine (VGT) valve during sudden closure of a butterfly valve. Detection of the surge limit in transient case is discussed.

## 2. Experimental set-up

### *2.1. Test bench description*

Experimental tests have been conducted on LIFSE test bench used to determine turbocharger characteristics [18]. A picture of the test bench is given in Figure 1.

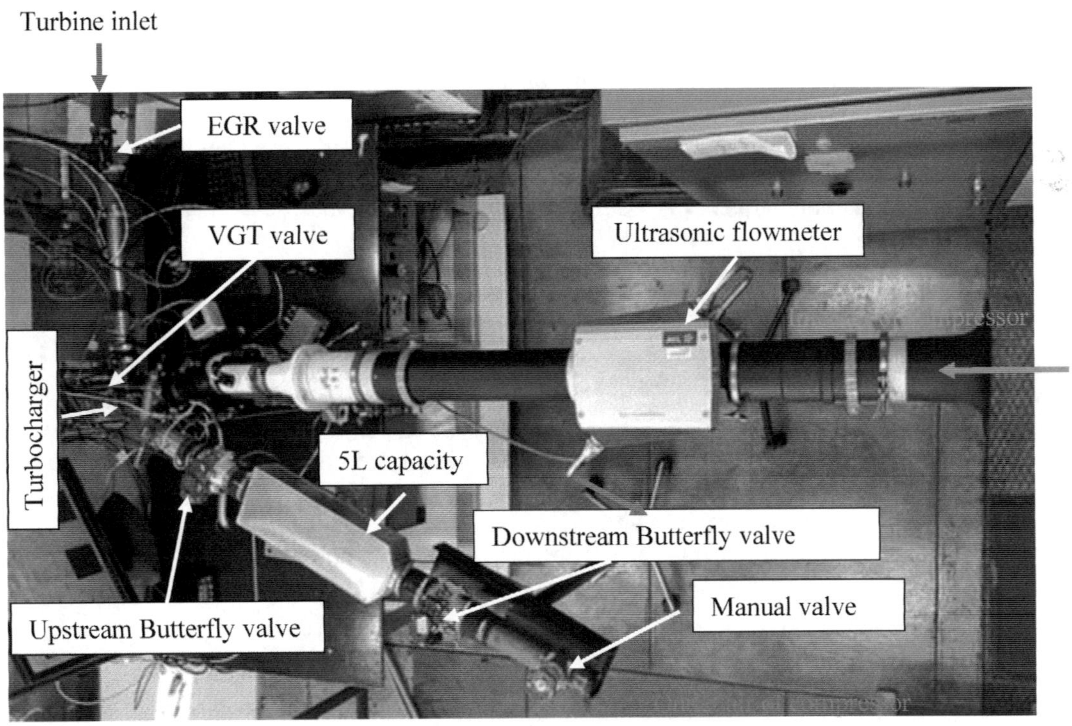


Figure 1. Experimental set-up in LIFSE laboratory.

Turbine is supplied with cold compressed air. Air flow rate is regulated upstream the EGR valve by a pneumatic valve. This EGR valve could be used to vary rapidly the flow rate. Compressed air is exhausted in the test room through a short duct added at the turbine outlet.

An ultrasonic AVL flowmeter is placed upstream the compressor to measure the instantaneous flow rate. An electrically controlled butterfly valve is placed downstream the compressor together with a 5L capacity, another butterfly valve and a manual valve. This manual valve is used to adjust the flow rate in the compressor while butterfly valves are used to modify rapidly the flow rate. The fast modification of flow rate is applied by the valves located downstream or upstream the capacity. The capacity is representative of a charge air cooler. This device amplifies the surge signal while the downstream valve is closed.

Measurements have been conducted on a MHI TDO3L turbocharger. The turbine body and its exhaust manifold are made of cast iron in one entire piece. This one is cut at the manifold position to allow compressed air to be supplied to the turbine.

### *2.2. Experimental measurements*

Piezoresistive sensors are used to measure pressure. Temperature measurements are made with PT 100 $\phi$ 3 mm 4 wires probes, the flow rate is measured with the AVL FlowSonic flowmeter, the flow rate of the turbine is measured with an orifice plate and a Optel optical sensor measures the speed rotation. These measurements allow to determine the characteristic curves of compressor and turbine in steady state regime.

Measurements of the temperature at compressor rotor inlet and at turbine volute inlet are acquired with K $\phi$ 3 mm thermocouples associated to a fast converter. Valve positions are given by voltage sensors.

All sensor signals are converted to 0-10 voltage and sent to a National Instruments USB 6218 acquisition card and recorded in an Excel file with a Labview program.

For measurements in steady state, the Labview program shows the operating point and its evolution in the compressor and turbine maps. This operating point is recorded when the stabilization is reached. For each operating point, 1024 data are acquired with a 500 Hz frequency.

For measurements in transient state, another Labview program is applied. Valves control and acquisition measurements use the same acquisition card and the same program. A specific « Producer-Consumer » type loop has been implemented to allow control of valves without impairing the data acquisition. Acquisition frequency is 1kHz and 5120 data are recorded allowing a test duration of 5s for one operating point.

This program permits to keep valves to a constant given position or to describe a "profile". For each valve, the profile is defined according to 6 time steps describing 6 positions.

### *2.3. Valves control*

Valves are actuated by a DC motor supplied with 12V. A sensor supplied with 0-5V determines the valve position. Valves control on engine is made by the ECU (Electronic Control Unit). Valves are supplied by a pulse width modulation (PWM) which is regulated according to the required position.

A UNO R3 Arduino microcontroller which can generate a 5V PWM signal is coupled to a motor module which can control valves with 12 V supply is used. A PID control governed by the microcontroller programing guarantee the closed-loop control position. The difficulty is to find the suitable coefficients to ensure a precise position control, by avoiding overshoot and oscillations and minimizing the response time.

#### *2.3.1. Butterfly valve control*

The butterfly valve, Figure 2, has a return spring which keeps the throttle valve open when the power supply shuts down. An example of obtained regulation is presented in Figure 3.

Figure 3 shows a slight difference between the control and the position. For a control fixed to 100%, the position is 97% and for 0%, it is 5%. This difference is inconsequential the valve can be considered completely open at 97% and closed at 5%". Flow rate variations at extreme positions have no significant effect on the flow rate through the valve.

Figure 2. Butterfly valve.

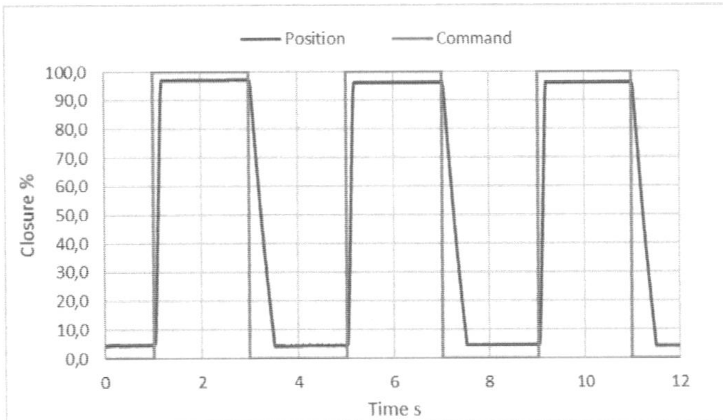


Figure 3. Butterfly valve response to a 0-5V step voltage.

The response time of closing is 0.155 s, of the same order of the response time obtained with the ECU. For opening, the response time is 0.535 s, mainly depending on the return spring as the voltage supply of the valve is shut down. The valve control using a given profile is satisfactory, especially since opening variations for the study are less large.

### *2.3.2. VGT control*

The VGT, Figure 4, actuator has no return spring. So, to inverse the direction of rotation, it is necessary to inverse voltage supply. The response of the valve to a 0-5V step voltage is represented in Figure 5. A very slight difference appears between the control and the valve position. The response time is about 0.2 s for opening and closing operations.

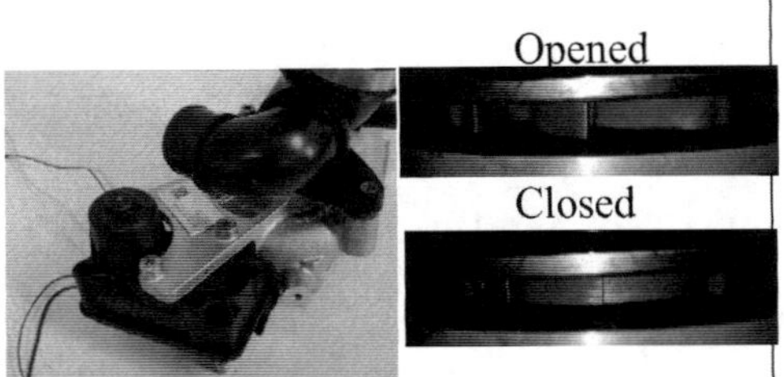


Figure 4. VGT valve.

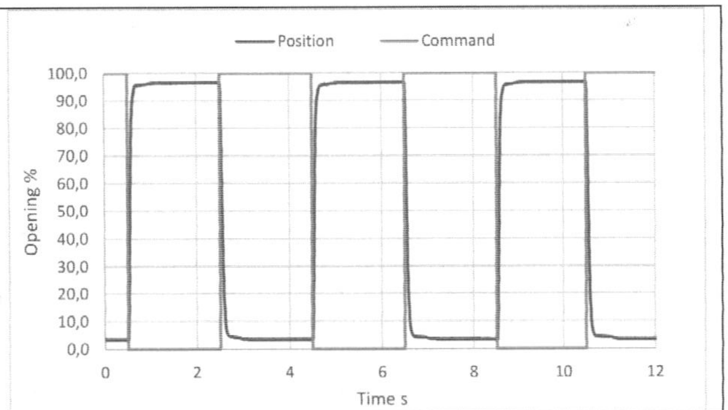


Figure 5. VGT response to a 0-5V step voltage.

## 3. Experimental measurements

### *3.1. Surge appearance detection*

Extensive tests were carried out for two rotational speed 101,000 rpm and 157,000 rpm. Only the tests with closing the valve downstream of the capacity at 101,000 rpm are presented here. Closure downstream gives high pressure fluctuations. For these tests, the adjustment of the initial operating point is obtained by closing the manual valve. In general, the surge limit is defined by the apparition of a low frequency peak (10-15 Hz) in the fast Fourier transformation (FFT) of the pressure signals while reducing the flow rate with the valve (mild surge). With the considered compressor in this study, this FFT peak does not appear, and deep surge is suddenly reach inducing important vibrations of the capacity. Figure 6 presents the instantaneous values of pressure ratio versus mass flow rate and the temperature measures at the compressor inlet close to the wheel $T_{wheel}$.

IOP Conf. Series: Earth and Environmental Science **960** (2022) 012013 doi:10.1088/1755-1315/960/1/012013

Pressure and flow rate fluctuations appear at point 6. They have the same magnitude for points 7 and 8. The brutal increase of temperature at rotor inlet $T_{wheel}$ indicate the presence of flow recirculation which is also a sign of rotating stall and surge. On this compressor it seems difficult to determine a clear criterion for the definition of the surge limit. This limit has been estimated for a mass flow rate of 0.02 kg/s for 101,000 rpm.

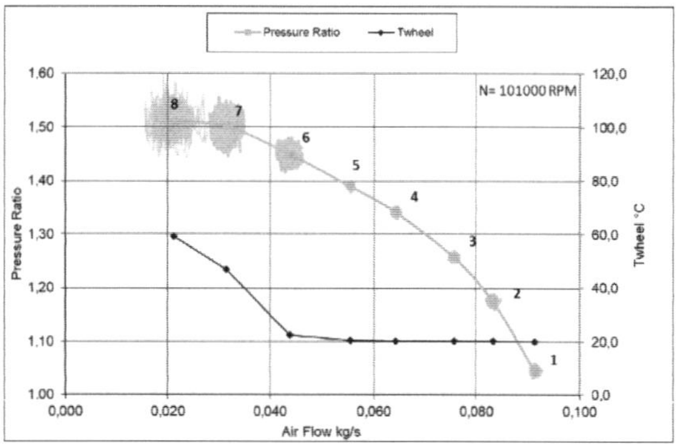

Figure 6. Pressure ratio and inlet temperature vs flow rate
Closure of downstream valve.

### 3.2. *Transient state experiments*

The experiments consist of positioning close to the surge limit point, using the stationary test program, and then with the transient test program, command the butterfly valve and VGT according to selected profiles. The initial position of VGT is chosen to obtain the minimum power to be supplied by the turbine (maximum efficiency of the turbocharger). Subsequently, only the results obtained for a speed of 101,000 rpm with butterfly valve downstream of the capacity are presented.

The manual valve is significantly closed, three control profiles are applied, Table 1.

**Table 1**. Butterfly valve closing profile.

| | | | | | | | |
|---|---|---|---|---|---|---|---|
| Profile 1 | Time s | 0.0 | 1.0 | 1,2 | 1,7 | 1,9 | 5.0 |
| (fast closing .2 s) | Closing % | 70 | 70 | 90 | 90 | 70 | 70 |
| Profile 2 | Time s | 0.0 | 1.0 | 1,5 | 2.0 | 2,5 | 5.0 |
| (moderate closing time .5 s) | Closing % | 70 | 70 | 90 | 90 | 70 | 70 |
| Profile 3 | Time s | 0.0 | 1.0 | 2.0 | 2.5 | 3,5 | 5.0 |
| (slow closing 1.0 s) | Closing % | 70 | 70 | 90 | 90 | 70 | 70 |

The VGT in its initial position is closed to 70%. A transient test is carried out in this position, Profile 0. A command to close to 90% in 0.2 s is then applied, simultaneously with the closing of the butterfly valve or in phase advance, see Table 2. The profiles have been defined according to a possible engine behavior control.

**Table 2.** VGT closing profile.

| | | | | | | | |
|---|---|---|---|---|---|---|---|
| Profile 0 | Time s | 0.0 | 1.0 | 1,2 | 3,2 | 3,5 | 5.0 |
| Constant position | Closing % | 70 | 70 | 70 | 70 | 70 | 70 |
| Profile 1 | Time s | 0.0 | 1.0 | 1,2 | 3,2 | 3,5 | 5.0 |
| delay 0 s | Closing % | 70 | 70 | 90.0 | 90.0 | 70 | 70 |
| Profile 2 | Time s | 0.0 | 0,8 | 1.0 | 3.0 | 3,3 | 90.0 |
| delay 0.2 s | Closing % | 70 | 70 | 90.0 | 90.0 | 70 | 70 |

| | | | | | | | |
|---|---|---|---|---|---|---|---|
| Profile 3 | Time s | 0.0 | 0.5 | 0.7 | 2.7 | 3.0 | 5.0 |
| delay 0.5 s | Closing % | 70 | 70 | 90.0 | 90.0 | 70 | 70 |

## 4. Results

Results for a fast closing of the butterfly downstream valve (Profile 1) and for the VGT Profiles 0 and 3 are presented in Figures 7.1 to 8.4.

Figures 7.1 and 8.1 show the initial position of the butterfly measured at 75% instead of the 70% that were set via the controller. This difference came from the positioning sensor calibration. It does not affect the results of the study and thus it was not considered necessary to apply a new calibration. The control of the VGT is good although there is an instability at 3.5s, which occur outside of the scope of the study.

In Figure 7.2, the increase of the rotational speed and the inlet temperature can be observed as expected due to the closing of the butterfly valve. In Figure 8-2, there is a clear drop in turbocharger speed when the VNT valve is closed and an increase when the butterfly valve is closed. However, the temperature increase occurs only after the butterfly valve closing. On these two graphs, the beginning of surge is indicated based on the temperature increase.

The closing of the VGT reduced the rotational speed and the outlet pressure P2c (Figure 8.3). The pressure fluctuations are less important than when the VNT stay at 70% (Figure 7.3) but they remain significant. The beginning of the surge estimated based on the analysis of P2c signal is the same.
Figure 7.4 and 8.4 show the simultaneous variation of the mass flow rate and the pressure ratio. The instantaneous mass flow rate is negative. The inversion of the flow is clearly a sign of deep surge operation.

Valve downstream the capacity
Profile 0 VGT – Profile 1 Butterfly valve - N = 101,000 rpm

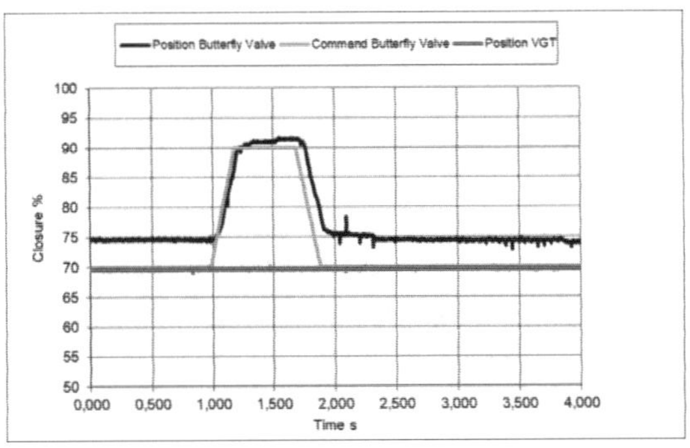


Figure 7-1. Valves control.

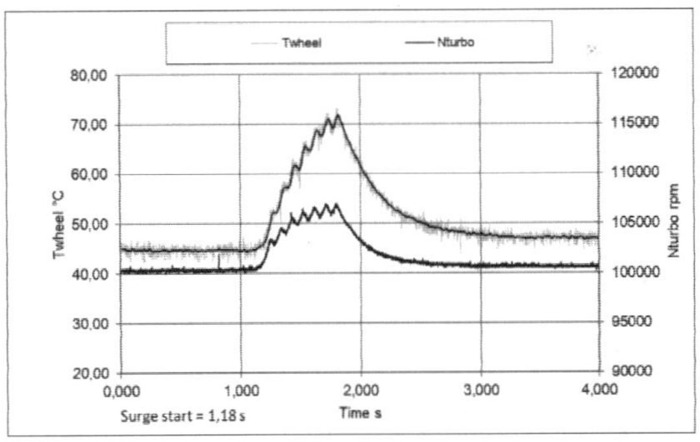


Figure 7-2. Inlet compressor temperature and turbocharger rotation speed.

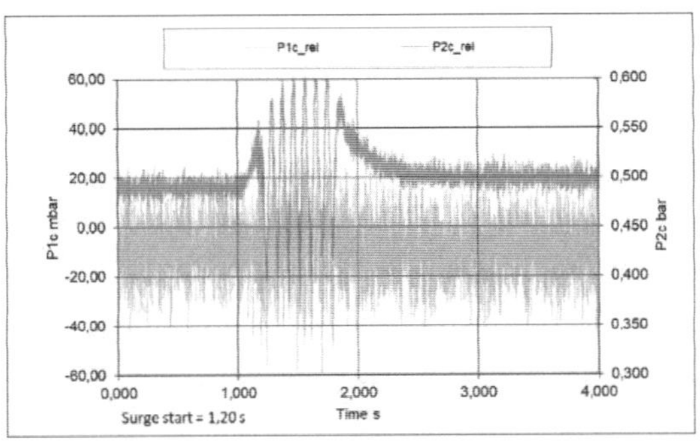


Figure 7-3. Compressor pressure inlet and outlet.

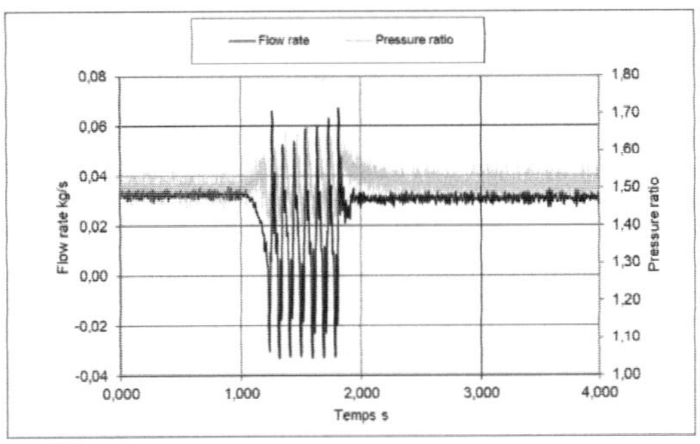


Figure 7-4. Flow rate and pressure ratio.

Valve downstream the capacity
Profile 3 VGT – Profile 1 Butterfly valve - N = 101,000 rpm

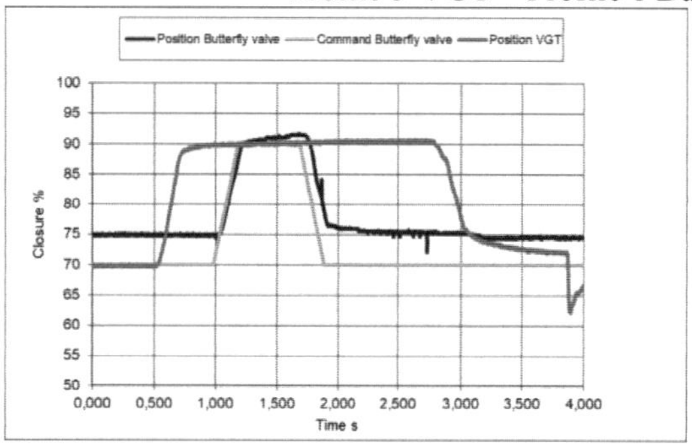


Figure 8-1. Valves control.

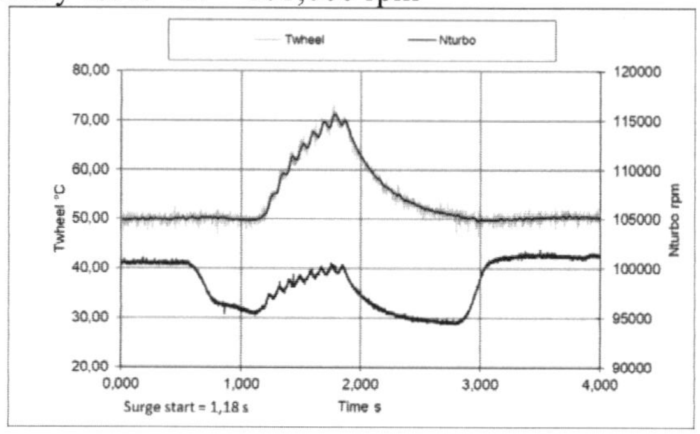


Figure 8-2. Inlet compressor temperature and turbocharger rotation speed.

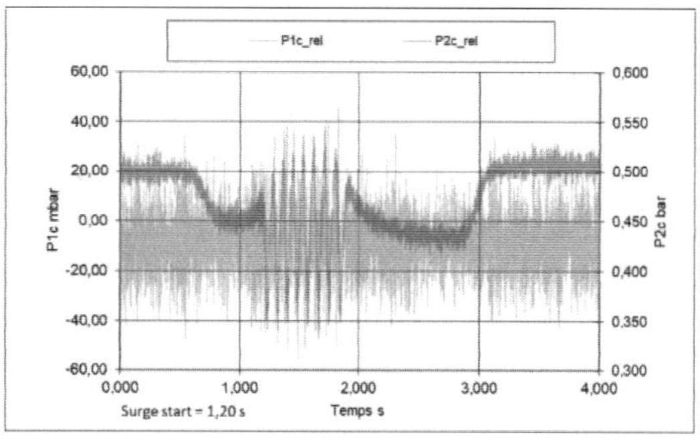


Figure 8-3. Compressor pressure inlet and outlet.

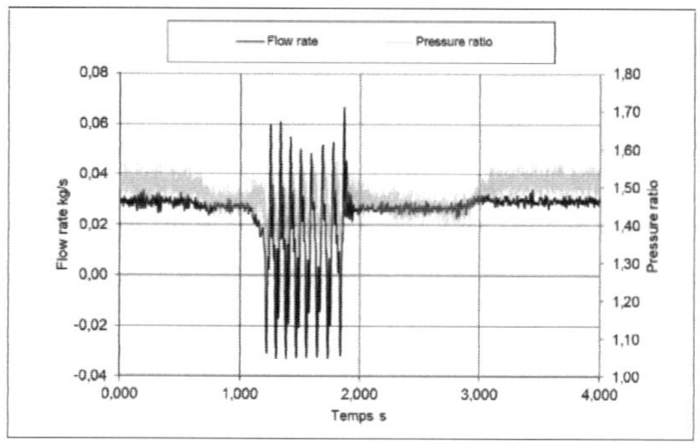


Figure 8-4. Flow rate and pressure ratio.

## 5. Conclusion

The experiments showed the difficulty to accurately define the surge limit of this particular compressor. The FFT analysis of the pressure and flow rate signals do not allow to observe the progressive apparition of the surge frequency. The surge appears suddenly and with even greater intensity when the butterfly valve is placed after the capacity. The temperature measurement at the rotor inlet highlights the presence of flow recirculation which occurs at flow rate relatively important compared to the surge limit. The analysis of instantaneous values of the pressure ratio versus flow rate shows significant oscillations with amplitudes comparable for stable operation and the surge limit. This makes the determination of the surge limit difficult.

Transient tests demonstrated that the amplitude of the fluctuations is stronger when the butterfly valve is downstream the capacity with significant variations of the rotational speed and the inlet temperature. This does not happen when the butterfly valve is directly placed after the compressor.

The closing speed of the butterfly valve has a logical influence on the beginning of the surge. However, it remains weak and difficult to evaluate precisely due to the impossibility to clearly set the beginning of the surge. Closing the VGT decrease the rotational speed, but it does not seem to have an influence on the beginning of the surge in both case when the VGT closing is synchronized or in advance with the butterfly valve closing. With the rotational speed reduction, the amplitude of the pressure ratio fluctuations decreases but the compressor enters into deep surge anyway according to mass flow rate inversion.

## Acknowledgements

We would like to thank Renault for funding this study and providing the material.

**References**

[1] Dong Kyu Kim, Jeong Hoon Seo, Seonyeob Kim, Min Kyu Lee, Ki Young Nam, Han Ho Song and Min Soo Kim, 2014 Efficiency improvement of a PEMFC system by applying a turbocharger, *Int. J. of Hydrog. Energy* **39** 20139-50

[2] Semlitsch B and Mihaescu M, 2016 Flow phenomena leading to surge in a centrifgual compressor, *Energy* **103** 572.

[3] Day IJ, 2016 Stall, Surge and 75 years of research, J. Turbomach. Trans. *ASME*, **138** 1-16

[4] Mohtar H 2010 Increasing surge margin of turbocharger centrifugal compressor automotive application, *Phd Thesis of Ecole Centrale de Nantes (ECN)*, France.

[5] Danlos A, Podevin P and Toussaint M 2017 Pre-swirl mechanism in front of a centrifugal compressor: effects on surge line and on unsteady phenomena in surge area. *MATEC Web of Conferences* **133**

[6] Greitzer EM, 1976 Surge and rotating stall in axial flow compressors. Part I: Theoretical compression system model, *J. Eng. Power* **98** 190-8

[7] Greitzer EM, 1976 Surge and rotating stall in axial flow compressors. Part II: Experimental results and comparison with theory, *J. Eng. Power* **98** 199-217

[8] Galindo J, Serrano J, Climent H, Tisira A, 2008 Experiments and modelling of surge in small centrifugal compressor for automotive engines, *Exp. Therm. Fluid Sci.* **32**(3) 818-26

[9] Marelli S, Carraro C, Moggia S and Capobianco M. 2016 Effect of circuit geometry on steady flow performance of an automotive turbocharger compressor, *Energy Procedia* **101** 630.

[10] Toussaint M, El Kamel C. 2012 Influence of the Network Geometry on the Surge Characteristics of a Centrifugal Compressor ISROMAC 14, Honolulu, United-States

[11] Wang X, Ge Y, Yu L, Feng X 2013 Effects of altitude on the thermal efficiency of a heavy-duty diesel engine, *Energy* **59** 543

[12] Galindo J, Tiseira A, Navarro R, Tari D, Meano C.M. 2017 Effect of the inlet geometry on performance, surge margin and noise emission of an automotive turbocharger compressor, *Appl. Therm. Eng.* **110** 875

[13] Bontempo R., Cardone M., Manna M and Vorraro G 2017 A statistical approach to the analysis of the surge phenomenon, *Energy* **124** 502.

[14] Andersen J, Lindström F and Westin F 2008 Surge definitions for radial compressors in automotive turbochargers, *SAE Int J Engines* **1** 218.

[15] Liu A.X. and Zheng X.Q. 2013 Methods of surge point judgment for compressor experiments *Exp. Therm. and Fluid Sci.* **51** 204.

[16] Galindo J, Serrano J.R, Guardiola C and Cervello C. 2006 Surge limit definition in a specific test bench for the characterization of automotive turbochargers, *Exp. Therm. And Fluid Sci.* **30** 449.

[17] Rakopoulos CD, Michos CN and Giakoumis EG 2005 Study of the transient behavior of turbocharged Diesel engines including compressor surging using a linearized quasi-steady analysis, *SAE Tech. Pap.*

[18] Deligant M, Danlos A, Podevin P, Clenci A and Guilain S 2017 Surge detection on an automotive turbocharger during transient phases, *IOP Conf. Ser.: Mater. Sci. Eng.* **252** 012082

IOP Conf. Series: Earth and Environmental Science **960** (2022) 012014 doi:10.1088/1755-1315/960/1/012014

# Assessment of the Miller cycle operation in a spark ignition engine via 1D numerical simulation

**P Punov[1*], M Niculae[2], A Clenci[2], S. Mihalkov[1], V Iorga-Siman[2], A Danlos[3]**

[1] Technical university of Sofia, 8, Kliment Ohridski blvd, 1756, Sofia, Bulgaria
[2] University of Pitesti, Str. Targul din Vale, nr. 110040 Pitesti, Arges, Romania
[3] Arts et Metiers Institute of Technology, Le Cnam, LIFSE, HESAM Université, F-75013 Paris, France

*plamen_punov@tu-sofia.bg ORCID ID: 0000-0003-4147-3755

**Abstract**. The article presents the results of a 1D numerical simulation of a spark ignition engine developed to operate in Miller cycle. Miller cycle offers better thermal efficiency compared to Otto cycle due to higher volumetric expansion than compression, which in the current context is of paramount importance. In an engine with fixed geometric compression ratio, Miller cycle operation could be realized by means of either early intake valve closing (EIVC) or late intake valve closing (LIVC). Both cases lead however to a lower volumetric efficiency, thus reducing the indicating mean effective pressure, which in its turn results to a lower power output. The simulation's aim is not only to assess the impact of implementing the Miller cycle but also to obtain the necessary results for imposing the boundary conditions in a 3D CFD simulation whose purpose is to analyse the influence of the Miller cycle on the internal aerodynamics of the engine.

## 1. Introduction

In order to achieve the global target of 55% reduction of greenhouse gases (GHG) by 2030 compared to 1990, European Union set a target of $CO_2$ emissions reduction from passenger cars and vans by 55% compared to 2021 and further reduction of 100% by 2035. In short term, internal combustion engines would be the main power source in transportation due to the higher vehicle autonomy, lower price and higher fuel energy density. For that reason, an improvement in thermal efficiency is important in order to respect the future $CO_2$ regulations. Over expanded engine operating cycle known as Miller-Atkinson cycle is a promising technology to increase engine efficiency and to reduce pollutant emissions [1,2]. The Miller cycle is based on late or early IVC, due to that it is easier to implement in both gasoline and diesel engines. The intake process could be also optimized when variable valve train (VVT) mechanism is used [3,4]. However, the optimization of gas-exchange processes and internal aerodynamics is important in order to achieve better engine performance and efficiency. The Miller operating cycle is widely studied numerically by means of 0D-1D and CFD simulations [3]. The validation of CFD modeling is usually done on a single cylinder optical engines operating at motored mode. However, in these experimental engines the blow-by could be higher as well as clearance volume due to crevices. Moreover, the blow-by increases at higher compression ration typical for Miller cycle. The study conducted on a motored engine in [5] revealed blow-by mass losses of nearly 15% at compression ratio of 12.4 with the maximum blow-by mass loss within the range of 6-9°CAD after top dead center. Another study [6] reported blow-by mass ratio within the range of 9% to 18% depending on compression ratio and engine speed. The results are also evaluated on a motored engine. The blow-by not only has significant impact on cylinder pressure but also it affects the peak position of the pressure so called loss angle. This effect was studied in [7].

The aim of this paper is to study the effect of blow-by on in-cylinder pressure of a single cylinder motored engine with high compression ratio developed to operate in Miller cycle. The blow-by mass flow will be further used as boundary condition in a 3D CFD model of the same engine in order to study the internal aerodynamics.

## 2. Engine simulating model

A 1D engine model on a single cylinder motored engine was developed in the advanced simulation software AVL Boost (Fig. 1). The pressure at the boundary sections of the intake and exhaust system were experimentally obtained, thus only the cylinder and valve ports were modeled with their geometric parameters. The main elements in the engine model are: general engine element (E1), cylinder (C1), plenums (PL1 and PL2), pipes (1 to 6) and system boundaries (SB1 and SB2).

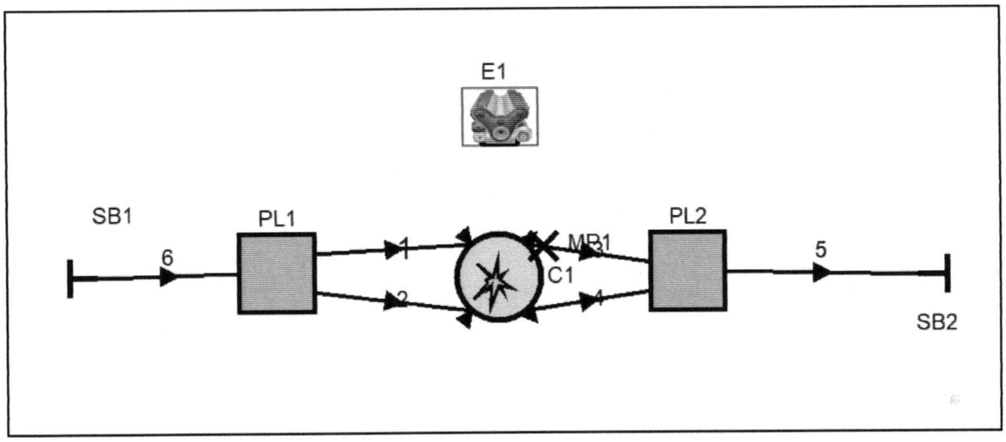


**Figure 1**. Engine simulation model in AVL Boost simulation software.

The main parameters such as engine speed and simulating cycle were defined in the element E1. Due to the fact that the simulation was done at motored mode (unfired operation), the mechanical losses in the engine were neglected. The main engine data are listed in Table 1.

**Table 1.** Engine data.

| Parameter | Value |
|---|---|
| Type | Mono-cylinder |
| Cylinder bore | 72.2 mm |
| Piston stroke | 70 mm |
| Volume | 286.4 cm$^3$ |
| Compression ratio | 12.3 |
| Valves per cylinder | 4 |
| Operating mode | Motored |
| Engine speed | 1200 rpm |

The element called "Cylinder" was used to define the geometrical parameters of the cylinder, the blow-by gap, valve ports and lift curves as well as heat transfer to the cylinder wall. The blow-by model is based on the following correlations:

$$Blow - by\, flow\, rate\ = \frac{dm_{bb}}{dt} = A_{eff} \cdot p_o \cdot \sqrt{\frac{2}{R_0 . T_0}} . \psi \tag{1}$$

where: $A_{eff}$ $[m^2]$ is the blow-by effective flow area; $p_0$ $[Pa]$ and $T_0[K]$ are the in-cylinder pressure and temperature; $R_0$ $[J/Kg.K]$ is the air constant and $\psi$ is the flow function. The latter is estimated as:

$$\psi = \sqrt{\frac{k}{k-1} \cdot \left[\left(\frac{p_2}{p_0}\right)^{\frac{2}{k}} - \left(\frac{p_2}{p_0}\right)^{\frac{k+1}{k}}\right]}, \tag{2}$$

where $p_2$ is the crankcase pressure, which was assumed to be a constant and equal to the atmospheric pressure and $k$ is the ratio between the specific heats.

The blow-by effective flow area is estimated as follows:

$$A_{eff} = D \cdot \pi \cdot \delta, \tag{3}$$

where $D[mm]$ is the piston diameter and $\delta[mm]$ is the gap between piston and cylinder.

Heat transfer was estimated with the commonly used Woschni model, presented in [8]. The model estimates wall heat transfer coefficient as a function of the cylinder bore, cylinder pressure and temperature as well as mean piston speed. Due to the motoring simulation, the temperature of the cylinder wall, piston top and cylinder head were assumed to be equal. The plenums "P1" and "P2" represents the junctions at cylinder head intake and exhaust ports. System boundaries were used in order to determine experimentally measured pressure at that points as a function of the crank angle.

The valve lift curves are developed for engine operating in Miller cycle. For that reason, the exhaust valve opening duration is higher while the intake valve maximum lift is lower. The relative valve lifts are presented in Figure 2.

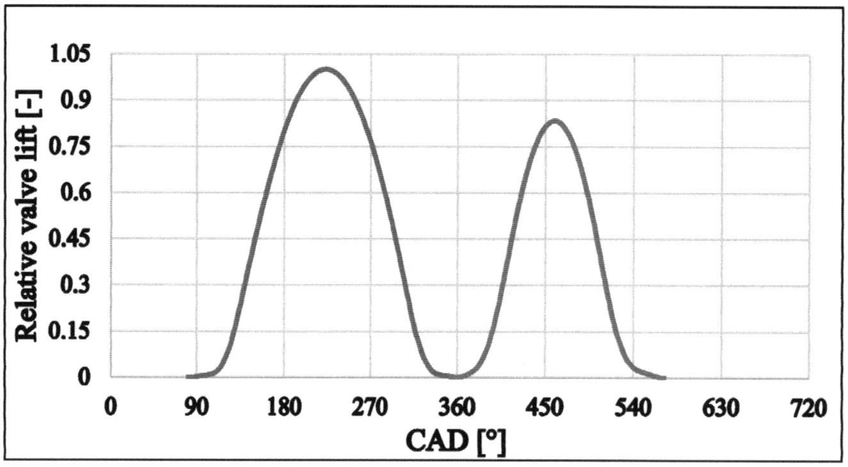


**Figure 2**. Valve lifts for operating in Miller cycle

## 3. Experimental data from engine motoring operation

In order to validate 1D simulation model as well as further validation of 3D CFD model an experimental study was conducted on a single cylinder optical engine. This engine was equipped with pressure transducers at the intake and exhaust cylinder head ports. Thus, further modelling of the whole intake and exhaust system geometry is not needed. The instantaneous pressure data was recorded for the studied operating points as a function of crank angle degrees (CAD) by increment of 1°CA. The experimental set-up can be seen in Figure 3.

**Figure 3**. Single cylinder optical engine

The experiments were carried out on engine operating in motoring mode at 1200 rpm. Four cases were studied as intake valve opening point was varied by 10 degrees at each study. Thus, in the first run the intake valve timing was set to the basic value as shown in Figure 2, then intake valve opening was advanced by 10, 20 and 30 degrees compared to basic valve timing. In general, earlier intake valve opening lead to lower in-cylinder pressure at the beginning of the intake stroke. The intake and exhaust pressure measured at cylinder head intake and exhaust ports for the last run (IVO advanced by 30 degrees) are presented in Figure 4. The TDC at the end of compression stroke corresponds to 360° CAD.

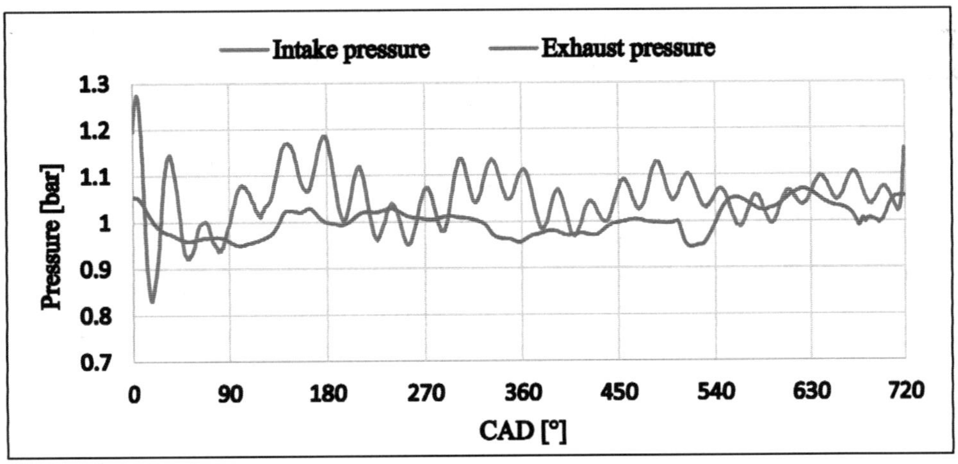


**Figure 4**. Experimentally measured intake and exhaust pressure

## 4. Simulating study

A 1D model was used in order to study the impact of both heat transfer and blow-by effect on in-cylinder pressure and loss angle. The blow-by effect was simulated by means of variation of blow-by gap in the model while the heat transfer effect was assessed by variation of heat transfer multiplier coefficient in the model.

### *4.1. Study the impact of in-cylinder heat transfer*

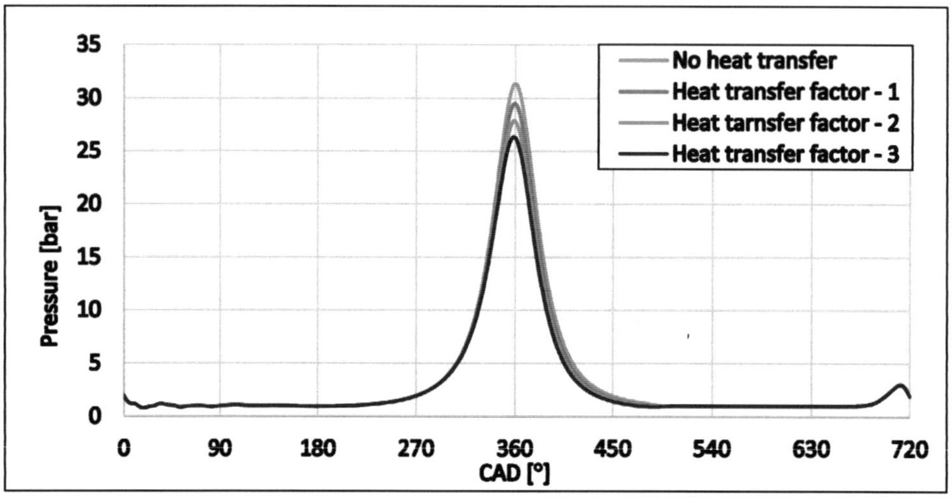


**Figure 5**. In-cylinder motoring pressure with different heat transfer rate

In order to modify in the model, the heat flux through the cylinder wall, piston top and cylinder head, the estimated heat transfer coefficient was multiplied by a coefficient for each case. It means that the instantaneous heat transfer value is multiplied by a coefficient. In this study the blow-by gap was set to 0 and the heat transfer multiplier coefficient varied from 0 to 3. The results concerning in-cylinder pressure are presented in Figure 5.

As the pressure and temperature has significant impact on the heat transfer coefficient, the impact on the in-cylinder pressure is important due to compression and expansion stroke. If no heat transfer is considered, the maximum in-cylinder pressure was estimated to be 31.3 bar as no loss angle was observed. Applying the heat transfer coefficient with factor 1, which means values estimated by original Woschni model were considered, reduced the maximum in-cylinder pressure to 29.5 bar as well as a loss angle of 0.6 CAD was observed. Increasing the heat transfer by multiplying factor 3 consistently reduced maximum in-cylinder pressure as well as the pressure during the expansion stroke. The loss angle was increased to 1.6 CAD. Due to the fact that the heat transfer in our case has no significant

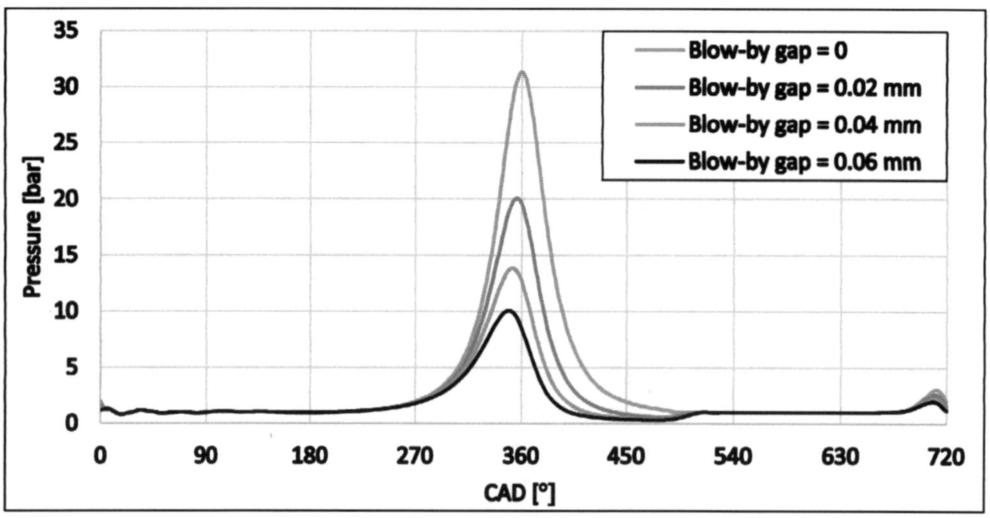


**Figure 6**. In-cylinder motoring pressure with different blow-by gap

IOP Conf. Series: Earth and Environmental Science **960** (2022) 012014 doi:10.1088/1755-1315/960/1/012014

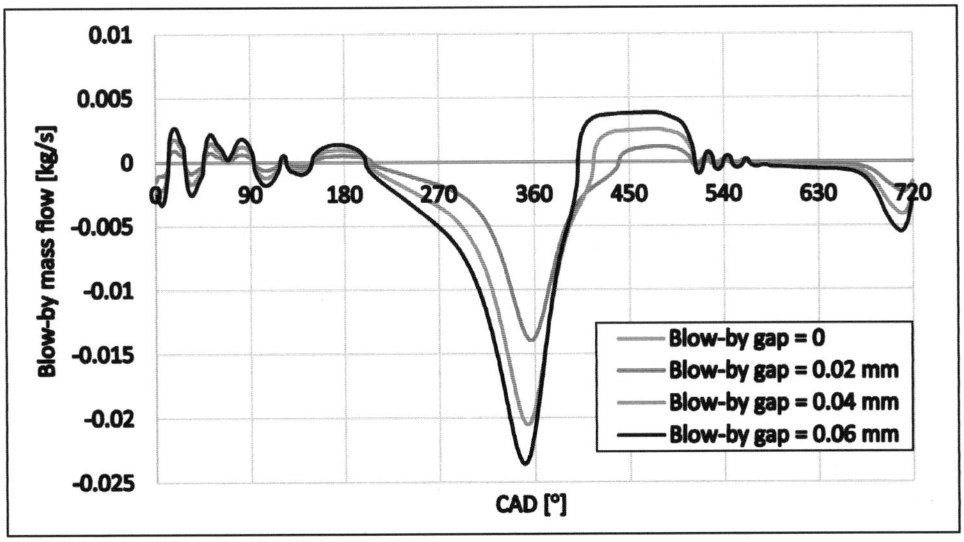


**Figure 7**. Blow-by mass flow with different blow-by gap

impact on gas exchange processes, we choose to use further original Woschni model with heat transfer factor equal to 1.

### *4.2. Study the impact of blow-by gap*

The blow-by mass flow and the impact on in-cylinder pressure was studied as the blow-by gap in the model was set to the value within the range of 0 mm to 0.06 mm. The crankcase pressure was set to constant value equal to the atmospheric pressure. By means of the simulating model, in-cylinder pressure was evaluated as well as the instantaneous blow-by mass flow as well as the intake and exhaust mass flow. Figure 6 shows the effect of blow-by on in-cylinder pressure. Increasing the blow-by to 0.02 mm reduced the maximum in-cylinder pressure to 20.1 bar compared to no blow-by case and loss angle was observed to be 4 deg. Further increasing of blow-by gap led to even lower maximum pressure. At blow-by gap of 0.06 mm the maximum in-cylinder pressure was reduced to 10.1 bar with loss angle of 8.5 deg. In order to better understand the impact of blow-by gap, the blow-by mass flow need to be analyzed. The results at different blow-by gaps are shown in Figure 7. The main cylinder discharge through blow-

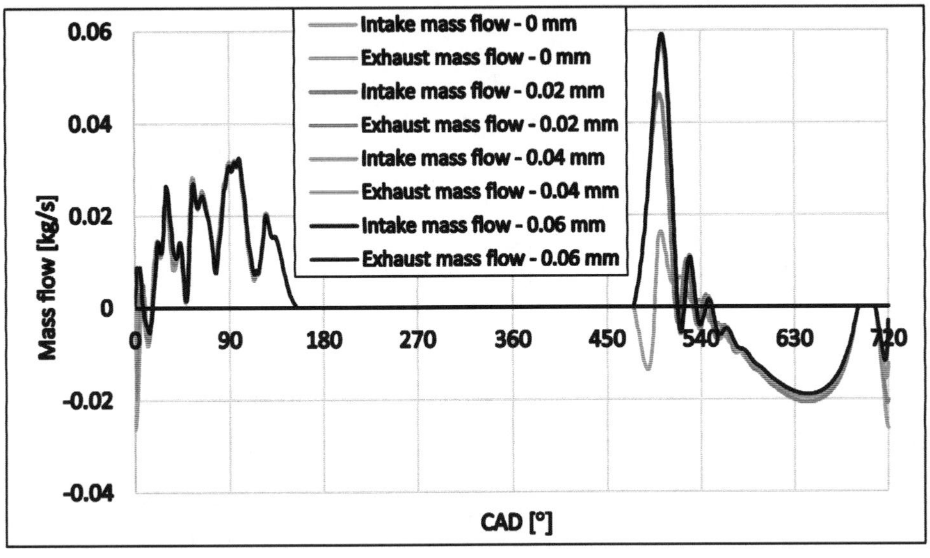


**Figure 8**. Intake and exhaust mass flow with different blow-by gap

by gap was observed during the compression stroke which led to reduction of in-cylinder mass. As a result, it lowers the in-cylinder pressure below atmospheric pressure during the expansion (Figure 6). Due to lowering the in-cylinder pressure below atmospheric pressure a significant cylinder charge through blow-by gap was observed. It became important at blow-by gaps higher than 0.02 mm. The blow-by gap has no important impact on blow-by mass flow during intake and exhaust stroke. However, at the end of exhaust higher blow-by gap led to higher blow-by mass flow when valve lift curves are developed for engine operating in Miller cycle.

Due to the significant impact on in-cylinder pressure during expansion stroke blow-by gap led to important cylinder backflow at the beginning of exhaust process. The mass flow through the valves is shown in Figure 8. It was observed that even small blow-by gap of 0.02 mm led to high backflow peak. However, increasing the blow-by gap to 0.06 mm did not lead to further increasing in the backflow compared to blow-by gap of 0.04 mm. Higher blow-by gap also led to higher cylinder discharged around TDC at the end of the exhaust stroke.

## 5. Conclusion

A numerical study of blow-by mass flow on a single cylinder motored engine operating under Miller cycle was presented. The model of the engine was developed by means of 1D simulation software AVL BOOST. Intake and exhaust pressure at the cylinder head ports were experimentally measured on a single cylinder optical engine at 1200 rpm. Firstly, the effect of heat transfer on in-cylinder pressure was discussed when using original Woschni heat transfer model and multiplying factor. Without heat transfer and blow-by the maximum in-cylinder pressure was estimated to be 31.3 bar. Increasing the heat transfer coefficient factor led to lower maximum pressure as well as to higher loos angle. Secondly, the impact of blow-by gap on in-cylinder processes was studied. For that study a heat transfer coefficient factor of 1 was chosen. Blow-by gap was varied from 0 to 0.06 mm. The results revealed that increasing the blow-by gap led to significant reduction of maximum in-cylinder pressure due to cylinder discharge during the compression stroke. It also lowering the in-cylinder pressure below atmospheric pressure during the expansion stroke. In case of 0.06 mm gap, the maximum pressure was observed to be 10.1 bar and the loss angle was estimated to be 8.5 deg. This higher blow-by mass flow led also to important cylinder charge at the beginning of exhaust valve opening. No significant impact was observed on an intake valve mass flow at different blow-by gaps.

The blow-by mass flow will be further used at 3D CFD model of the same single cylinder engine operating in the same set-up in Miller cycle.

## Acknowledgment

The study was funded by project "Passenger cars efficiency improvement by means of optimized operation engine-turbocharger" № КП-06-Франкофония/3 within the frame of bilateral program BNSF-AUF session 2018 and supported by the project BG05M2OP001-1.001-0008 "National Centre for Mechatronics and Clean Technologies" Operational Program Executive Agency, Ministry of Education and Science, Bulgaria. Three partners were involved: Technical University of Sofia, Bulgaria, Cnam Paris, France, and University of Pitești, Romania.

The numerical simulations in the study were done by means of the advanced simulation package AVL AST. We gratefully acknowledge the AVL company for providing us with the opportunity to use AVL simulation products for numerical studies at the Faculty of Transport of the Technical University of Sofia.

## References

[1] Zhao J 2017 Research and application of over-expansion cycle (Atkinson and Miller) engines – A review *Appl. Energy* **185** 300–19
[2] Wei H, Shao A, Hua J, Zhou L and Feng D 2018 Effects of applying a Miller cycle with split injection on engine performance and knock resistance in a downsized gasoline engine *Fuel* **214** 98–107

[3] Perceau M, Guibert P and Guilain S 2021 Zero-dimensional turbulence modeling of a spark ignition engine in a Miller cycle « Dethrottling » approach using a variable valve timing system *Appl. Therm. Eng.* **199** 117535
[4] Wang J, Duan X, Wang W, Guan J, Li Y and Liu J 2021 Effects of the continuous variable valve lift system and Miller cycle strategy on the performance behavior of the lean-burn natural gas spark ignition engine *Fuel* **297** 120762
[5] Aghdam E A and Kabir M M 2010 Validation of a blowby model using experimental results in motoring condition with the change of compression ratio and engine speed *Exp. Therm. Fluid Sci.* **34** 197–209
[6] Irimescu A, Tornatore C, Marchitto L and Merola S S 2013 Compression ratio and blow-by rates estimation based on motored pressure trace analysis for an optical spark ignition engine *Appl. Therm. Eng.* **61** 101–9
[7] Pipitone E and Beccari A 2010 Determination of TDC in internal combustion engines by a newly developed thermodynamic approach *Appl. Therm. Eng.* **30** 1914–26
[8] Woschni G 1967 A Universally Applicable Equation for the Instantaneous Heat Transfer Coefficient in Internal Combustion Engines *SAE 6700931*

# Domestic peat pots biodegradation

**F Uleanu*, D Giosanu, M Vulpe, G Bratu and L Vîjan**
University of Piteşti, Str. Târgu din Vale, no. 1, 110040, Piteşti, România

E-mail: uleanuflorina@yahoo.ro
ORCID: https://orcid.org/0000-0003-4825-7596

**Abstract.** Biodegradable nutrient media based on cellulose fibers and peat with added protective materials and stimulating represent the higher form of transplant used in current technologies for producing plant seedlings. Worldwide intensive development of the pots production industry (the effect of increasing demands from agriculture and their quality level in terms of degradability and operational sustainability, with outstanding results in reducing stress factors) led to fast expansion of their use with a high economic efficiency. Determination of potential and effective degradation was achieved by the method for determining cellulolytic activity in soil. Biodegradation potential recorded differences during the course of experience in the sense that it increases as the pots exposure time to treatment is higher. In conclusion, the pots treatment (planted/ without plant) has a great influence on the effective biodegradation.

## 1. Introduction

The choice of systems for obtaining vegetable seedlings in the world offering a wide enough range, adapted the needs of small farmers and large agricultural associations. Conventionally vegetable crops practiced to high technological standards (EU), provides highly stimulating profits reflected an enviable standard of living.

By using different types of pots:
- increase the flexibility of to establish periods of culture,
- it protects and insulates plant injuries or damage due to handling and transport,
- is protecting the roots from drying during storage and transport,
- facilitate adaptation of mechanization seedling production,
- is achieved vegetable crops with optimal density and an early maturing crop.

The negative impact on the environment of plastic pots used in agriculture was reduced by using the biodegradable pots.

## 2. Material and methods

In order to achieve the objectives we used the following materials:
- Pot compositional recipe:
    - Poiana Stampei peat 70%
    - secondary cellulose fiber 30%
    - Kymene resin 611 (commercial product) 3%
    - chitosan 1%
    - auxiliary chemicals (solution 3%)2 %
- Terracult peat;

- Tomato seedlings, variety Rila F1.

Were mounted two experiences monofactorial type:

1. Evaluation of biodegradation potential of pots made on the recipe produced with the pilot plant for biodegradable nutrient media training, designed and built by Ceprohart Brăila SA.

   Experimental variants, represented by the sampling times were as follows:
   - T1- 1 collection time, after 3 weeks of treatment,
   - T2- 2 collection time, after 6 weeks of treatment,
   - T3- 3 collection time, after 7 weeks of treatment.

2. Determining the effective state of degradation, technological operating conditions to produce tomato seedlings, the variants obtained were the following:
   - V1- pots filled with nutrient substrate, but without seedlings,
   - V2- pots with tomato seedlings.

To assess the potential for biodegradation of pots, they were weighed, coated in synthetic, no biodegradable lycra type and planted in plastic pots with sides of 12 cm, filled with Terracult nutrient substrate. Periodically amounted test sample, so variants are represented by the sampling times. Experimental period was about 2 months. Throughout this period, effective humidity of the substrate in which the pots were buried was maintained at about 55%.

To determine the effective state of degradation, each pot was weighed dressed in lycra type no biodegradable material and reweighing, filled with Terracult nutrient substrate and planted with tomato seedlings. Throughout the experimental period were applied uniformly seedling care works in accordance with the specific technology (daily ventilation, watering, weeding weeds).

Sampling analysis was done when tomato seedling reached the optimum planting stage.

## 3. Results and discussions

In terms of biodegradation potential - experience1, differentiation occurs in the sense that it increases as the pots exposure time to treatment is higher.

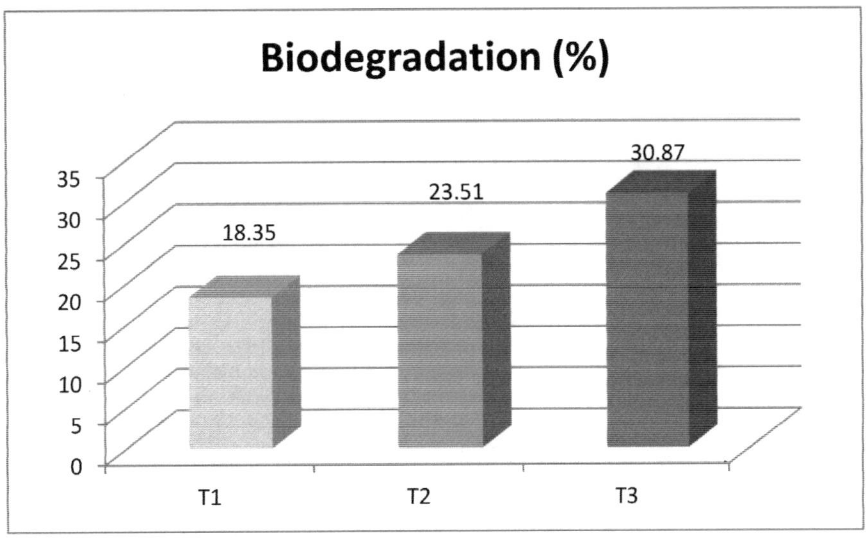


**Figure 1.** Biodegradation potential (%).

At 21 days the biodegradation potential was 18.35%, at 42 days of 23.51% and 30.87% at 49 days (figure 1).

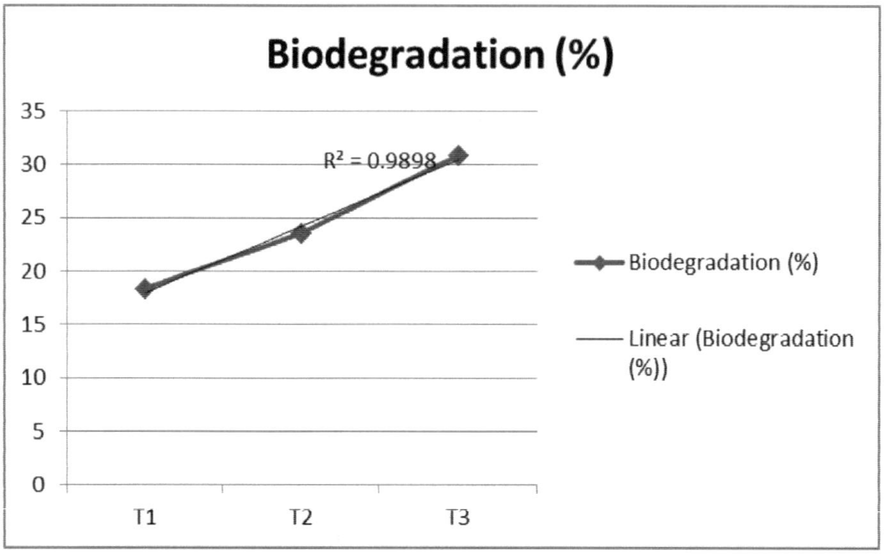


**Figure 2.** Influence of duration of exposure time
on the overall rate of biodegradation of pots without plants - experience 1.

The duration of exposure to treatment influence biodegradation potential significantly distinct (total rate of biodegradation) and average daily rate of biodegradation.

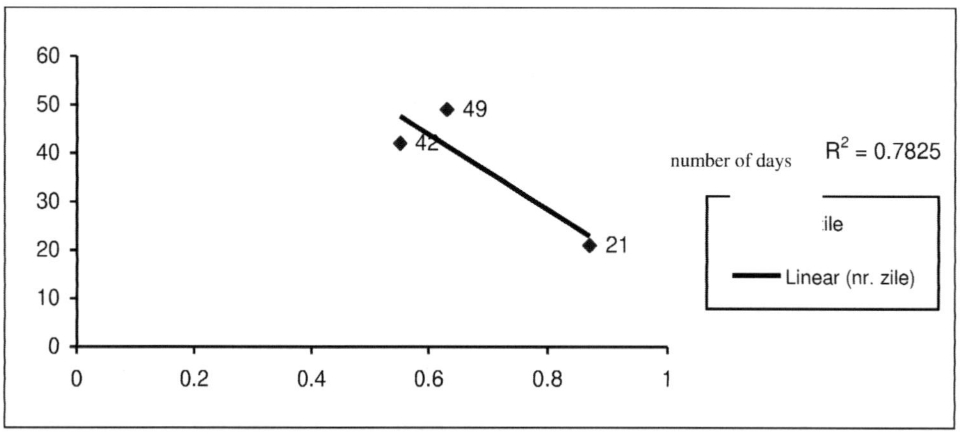


**Figure 3.** Influence of duration of exposure time
on the average daily rate of biodegradation of pots without plants - experience 1.

**Average daily rate of biodegradation** also presents differences related to pots treatment, registering higher values on pots without plants (0.46%/ day) compared to the pots with tomato seedlings (0.34%/ day). Average daily rate of biodegradation is highest (0.87% / day) at the first moment of analysis, the lowest value of this indicator recorded in the mid-range of analysis, namely 42 days.

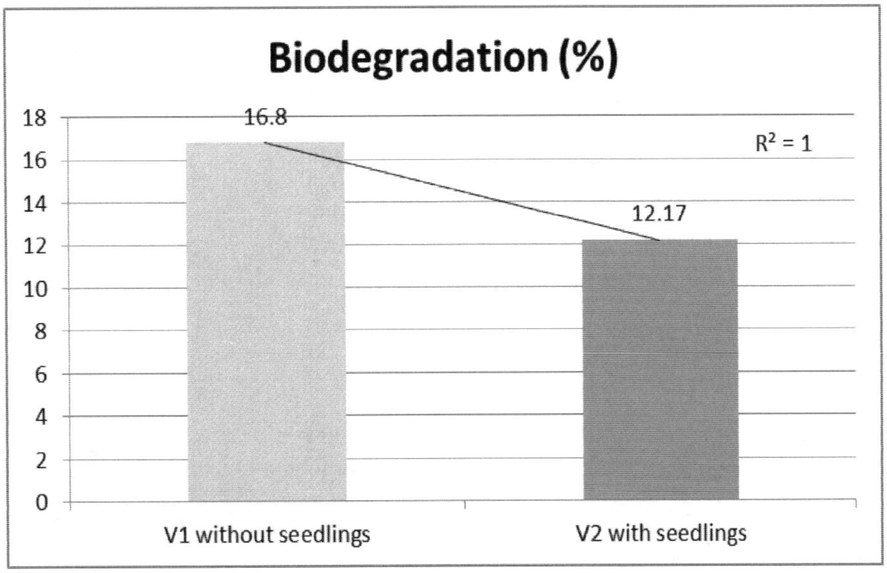


**Figure 4.** Treatment influence on the effective pots biodegradation.

In terms of effective biodegradation of pots - experience 2, shows that at the pots without plants (V1) was a higher effective rate of biodegradation, 16.80%, than in V2 tomato seedlings 12.17%.
It found that the pots treatment (planted / without plant) has a great influence on the effective biodegradation.

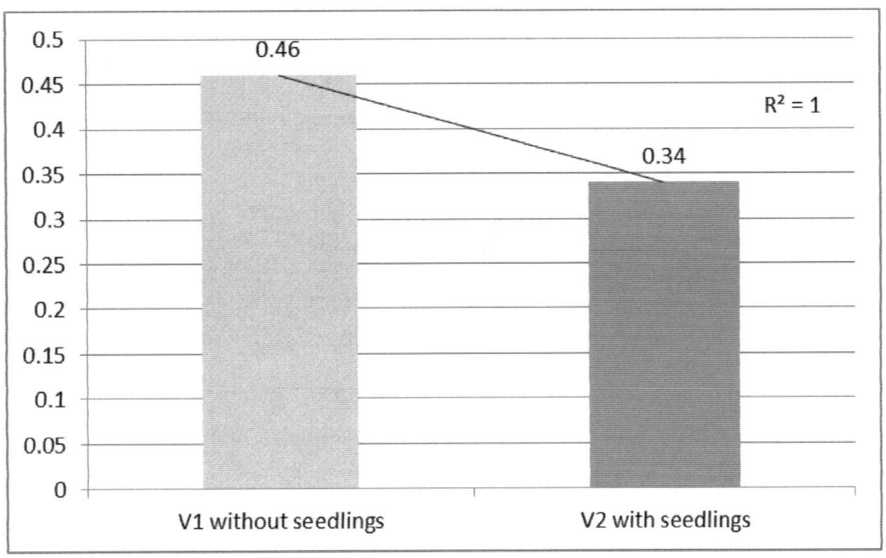


**Figure 5.** Influence of treatment of the average daily rate of pots biodegradation - experience 2.

On the whole experience, it appears that the presence of plants, by the rhizosphere effect, causes a decrease of 4.63% effective biodegradation. Also, as a result of the treatment, is strongly influenced daily rate of biodegradation, the presence of plants in pots, causing a decrease in this indicator of 0.12%/ day (Figure 5).

## 4. Conclusions

Biodegradation potential recorded differences during the course of experience in the sense that it increases as the pots exposure time to treatment is higher.

The duration of exposure to treatment influence significantly distinct biodegradation potential (total rate of biodegradation) and average daily rate of biodegradation.

More generally the two experiments conducted in parallel, we find that on average, in operating conditions, biodegradation is 59.77% lower than the potential for biodegradation.

This can be explained by the fact that plants, for growth and development consumes a lot of nutrients, competing for food microorganisms involved in biodegradation.

## 5. Bibliography

[1] Epure V, Nechita P and Dobrin E 2007 Utilizarea fibrelor celulozice secundare în amestec cu turba de suprafaţă la obţinerea suporturilor nutritive biodegradabile, *Celuloză şi hârtie* **56(4)** 33

[2] Epure V, Nechita P and Dobrin E 2008 Suporturi nutritive biodegradabile din resurse regenerabile pentru producerea containerizată a materialului săditor, Simpozionului ştiinţific anual cu participare internaţională "Horticultura – ştiinţă, calitate, diversitate şi armonie", Iaşi, p 57

[3] Epure V, Nechita P and Dobrin E 2008 Composite Materials Based on Lignocelulosic Fibres Used as Nutritive Supports in Seedling Manufacturing, 3rd International Conference & Exhibition on Environmentally Compatible Forest Products ECOWOOD Porto – Portugalia, p 350

[4] Ciofu R, Nistor S, Popescu V, Chilom P, Apahidean S, Horgoş A, Berar V, Lauer K F, Atanasiu N 2003 *Tratat de legumicultură*. Editura Ceres, Bucureşti, p 251

[5] Davidescu V, Costea G, Madjar R, Stanica F and Caretu G 2002 *Substraturi de cultură*. Editura Ceres, Bucureşti, p 136

[6] Davidescu V, Madjar R and Mănescu C 2008 *Valorificarea unor deşeuri reciclabile în cultura containerizată a unor specii dendrologice*, Editura Invel Multimedia, Bucureşti, p 114

[7] Pawlik M, Cania B, Thijs S, Vangronsveld J and Piotrowska-Seget Z 2017 Hydrocarbon degradation potential and plant growth-promoting activity of culturable endophytic bacteria of Lotus corniculatus and Oenothera biennis from a long-term polluted site. *Environ Sci Pollut Res* **24** 19640-52

[8] Compant S, Duffy B, Nowak J, Clement C and Barka EA 2005 Use of plant growth-promoting bacteria for biocontrol of plant diseases: principles, mechanisms of action, and future prospects. *Appl Environ Microbiol* **71(9)** 4951-9

[9] Treinytea J, Grazulevicienea V, Bridziuviene D and Svediene J 2014 Properties and behaviour of starch and rapeseed cake based composites in horticultural applications. *Estonian Journal of Ecology* **63(1)** 15-27

# Experimental research concerning the possibility of the cow manure conversion into biogas

**I A Halmaciu**[0000-0002-9932-7366]**, I Ionel**[0000-0002-2999-1210] **and I Vetres**[0000-0002-6425-9818]

[1]Politehnica University of Timisoara, Romania, Faculty of Mechanics, Bv. M. Viteazu, No. 1, 300222, Timisoara, Romania

E-mail: ioana.ionel@upt.ro

**Abstract.** Cow manure has been used since ancient times as an organic fertilizer in agriculture. Being rich in micronutrients, especially zinc, it is suitable to be used for all types of crops. The increase in the number of inhabitants generated the need for a larger amount of food and associated ever growing need of energy supply. Thus, an increase in the number of farms and cultivated agricultural areas was necessary, as well ever-growing secure energy sources. The strict laws on waste management, on reduction of greenhouse gases emission and in general addressing the decrease of the manmade pollution are some of the reasons that have led to the developing and application of alternative energy solutions. This paper aims to shed light on another much more beneficial feature of cow dung through its modern use, by production of biogas through anaerobic digestion. The experimental work is performed on an ultra-modern laboratory installation, capable to generate in 6 separate digesters of the biogas, according todivers receipts. In order to create a favourable environment for the growth of bacteria, the digestate was kept at a temperature of 37° C, thus creating mesophilic conditions. Throughout the experiment the pH value was monitored and recorded, as well the amount of generated methane.

## 1. Introduction

Since early times, the main preoccupation of people has been breeding animals and cultivating the land. In order to obtain a bigger production of cereals, the zoo-technical wastes were used as organic fattening [1].

The big number of the global population in the last years has generated the necessity of a big food quantity, which led to the cultivation of increasing areas of arable land and to increasing number of animals.

If in the past, it was sufficient for the animal farms to function only for a profit, nowadays these must function efficiently, both from an economic point of view and as well an ecological one [2].

The tendency of the increasing of the animal numbers causes a significant quantity of zoo technical wastes for which there should be ensured special storage spaces, and then urgently finding solutions of valuing/eliminating with a reduced impact on the environment.

Owing to the herbivore food of horned cattle, the waste that they produce contains a high percentage of fibres, and it is also rich in micronutrients, especially Zinc. All these proprieties make it suitable to all types of crops [3,4].The cow manure contains the following elements: organic humus (50-70%), nitrogen (1.8- 2.4%), phosphorus (1.0-1.2%), potassium (0.6-0.8%), and in small quantities there can be found magnesium, zinc, iron [5].The direct application of cow waste on the soil does not represent only a source of gas emissions with greenhouse effect, but this fact can lead to contamination of soil and the underground water[6,7].The salts, the bacteria, the viruses, which are found in the excrements, once reaching the soil and the water, can become a real danger for the human health [8].

The laws that become harsher and harsher concerning the manure managing (the law approval: the Law no. 123 from 10 July2020 for the changing and the completion of the urgent written order of the Government no. 195/2005 concerning the environment protection, known as "The Smell Law"), the gases with the greenhouse effect [9], and the people`s activity, which need an important quantity of energy [10], represent only a few of the reasons for finding and applying alternative energy solutions. A solution that must not be neglected, and seriously must be taken into account, is the biogas production. The transforming of the cow manure energy content, with the help of an installation of biogas production, into electrical and heat energy could significantly reduce the ecological damage caused by the improper elimination of the waste [11]. Thus, the anaerobe digestion (AD) can be considered as one of the most appropriate solution to solve the problem of the zootechnical wastes [12].

Due to the high content of fibres in the cattle manure there is a greater percentage of energy making it much more valuable in comparison with the other types of zootechnical wastes to be used as raw material for biogas production [3,13].

Another major problem is represented by the unpleasant smell caused by the zootechnical installations, which can be solved by using the biogas installation [14].

The remaining parts after the energy extraction are changed into digested paste which does not have a significant energy value anymore [15,6]. This paste is known as the digestate or compost and it can be used as fattening for agricultural lands, as it contains nutritious substances such as: nitrogen, phosphor, potassium.

The studies have proved the fact that the digestate has the capacity to rapidly release the nitrogen in the soil this fact brings benefits to the plants [16]. By the application of the digestate on the agricultural lands there is an improvement of the soil quality, and the chemical fattening amount is reduced. Due to the great number of proprieties, one can state that the digestate (compost) can be used as soil balm for the stimulation of plant growth [17].

For a biogas installation to be efficient for a farm one must take into consideration the following aspects: the quantity produced by the waste, the water content in the cow manure and its potential to produce biogas [18].

The installation of biogas production creates a lot of advantages, not only regarding the environment protection by reducing the soil and water pollution, but as well as from an economic point of view by encouraging the circular economy [19]. These benefits have recently led to being given a global attention to [20].

## 2. Materials and methods

### *2.1. The quantity and the raw material used*

In order to achieve this experiment, it was used the cow manure from a mini cattle farm and the fresh inoculums from a biogas production facility, all these being mixed with warm water. The inoculums were kept "for starving", over 7 days, at a temperature of about 37°C, to maintain fresh their microbial activity. Throughout the whole experiment the organic fraction was kept in a fridge.

The experiment was carried out on a laboratory device made of 6 digesters with a volume of 2000 ml, but during the experiment the loading rate was of 1800 ml.

For a start, each of the 6 digesters was loaded with: 905.85 g inoculums, 404.42 g of cow manure and 489.73 g of water. During the whole experiment the bacteria from each digester were daily fed at the same hour with 66.1 g of cow manure and 80.0 g of water. Before every supplying process there were extracted 146.1 g of digestate from each digester.

### *2.2. The method of obtaining the biogas*

The process of obtaining the biogas was mesofilique, where the temperature can vary between 25-40°C [21]. In order to maintain this regime, the 6 digesters were kept in a thermostatic water bath, whose temperature was set at 37°C. Taking into account the fact that the pHcan directly affect the microorganisms [22], with the help of a pH-meter. Thus, digestate pH measurements were made daily.

Considering that temperature plays an important role in the production of biogas, the digestate was also supervised and from the point of view of the temperature evolution.

To obtain a large quantity of biogas it must be taken into account the hydraulic retention (HRT) and of the organic loading rate (OLR) [23].The hydraulic retention time (HRT) shows the duration that a material can spend in the digester [24], while the organic loading rate (OLR) indicates the quantity of the stratum used for supplying the digester every day [25].

The homogeneity of the content inside the digesters was achieved with the help of the agitators, which are part of the biogas installation. In order to maintain a homogenization and a good density, the functioning time of the agitators was set at 10 seconds, while the stopping period of the agitators was set at 50 seconds. The agitators have the role to prevent the forming of the crust in the digesters [26].

*2.3. Results and discussions*

The laboratory installation being a very high performance one, during the whole experiment (28 days) there could be observed the gas flow, the temperature, and the specific production of the gas.

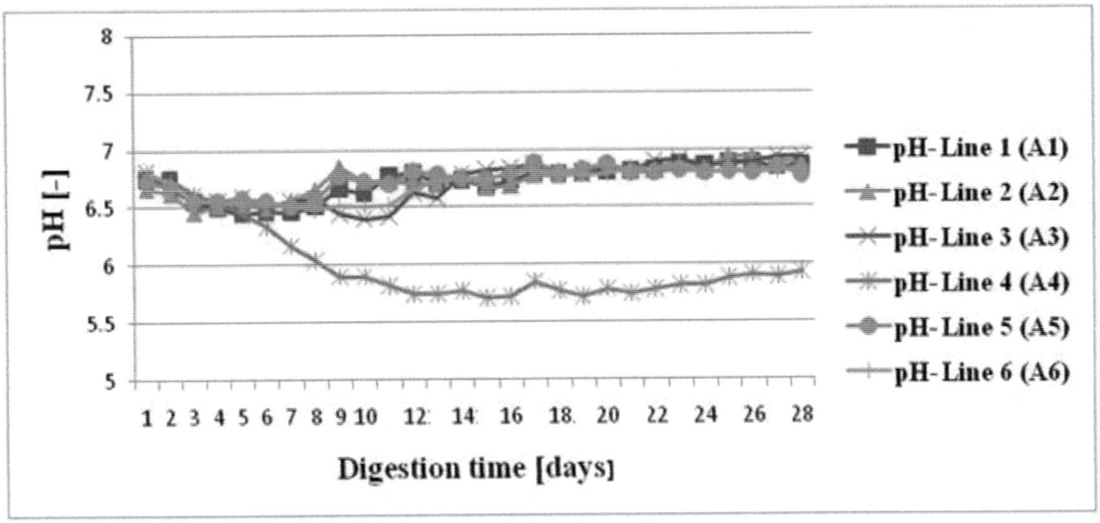


**Figure 1**. The recorded pH evolution in the 6 digesters

Figure 1 shows the values of the pH recorded in the 6 digesters, within the 28 days of the experiment. In the digesters A1, A2, A3, A5 and A6 the pH evolution varies between 6.40 and 6.94. For A4 the (A4), due to the improper supplying of the digester in the first days of the experiment, the value of the pH values was smaller (between 5.71 and 6.76). Thus, one did not take into account the analysed activity in digester4, caused by the fact that the phenomenon was not developed/initiated/ properly.

The temperature variation for each of the 5 lines is represented in figure 2. Throughout the whole experiment the temperature was relatively constant, sometimes being recorded small variations of ± 2° C. This fact proves that the bacteria were created an optimal developing environment in a mesofilique regime.

In figure 3 it is presented the gas flow rate recorded, using the cow manure as raw material. From this graphic one can observe that in the first days of the experiment the gas flow rate was more reduced, managed to record a significant increase especially at line 2 (A2) starting with the 10th day. During the experiment the reoccurred decreases off gas flow rate at all the lines, but the most significant decreases were recorded at line 3 (A3). These decreases triggered relatively quickly, after the gas flow at this line began to increase. The days when the gas flow rate was reduced are as follows: 17, 18, 19 and 20. The largest gas flow rate was recorded at line 2 (A2), during the $10^{th}$ day, while the smallest level of the flow was recorded at line 3 (A3), in the $17^{th}$ day.

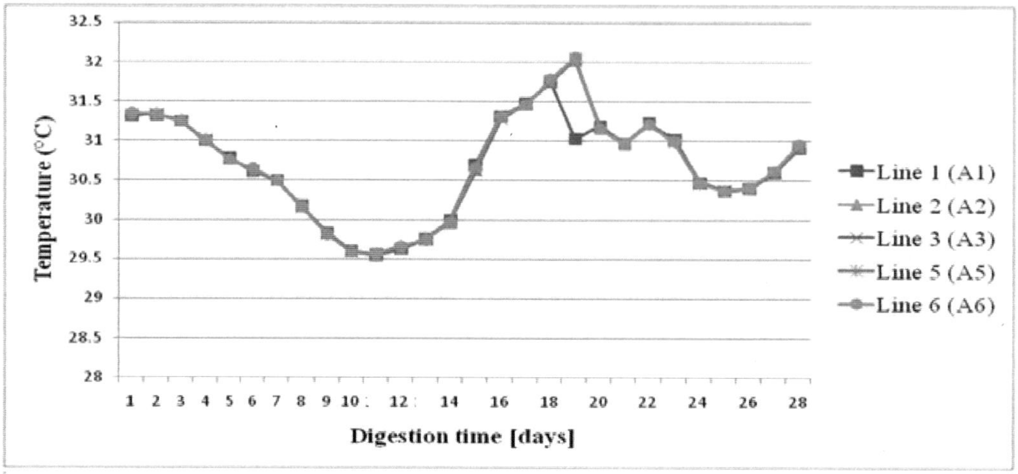


**Figure 2.** The temperature variation on days for the 5 lines

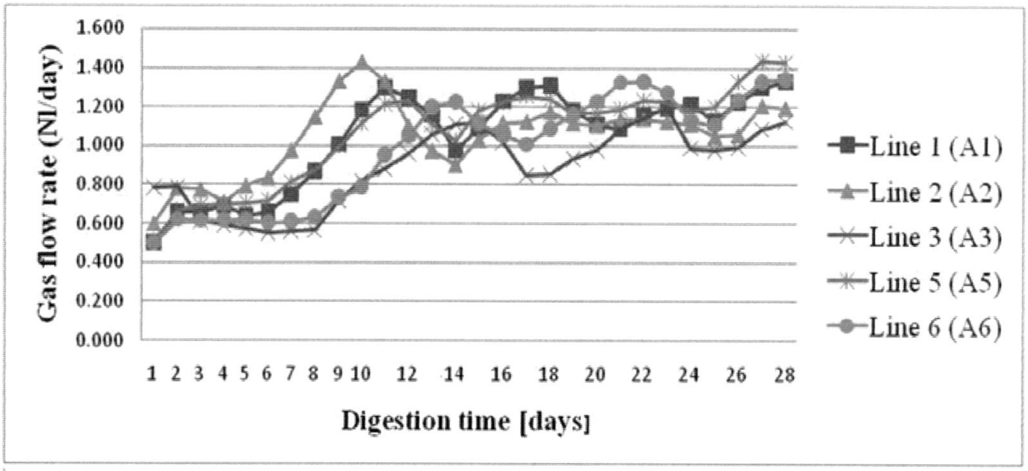


**Figure 3.** The recorded flow rate of the produced gas

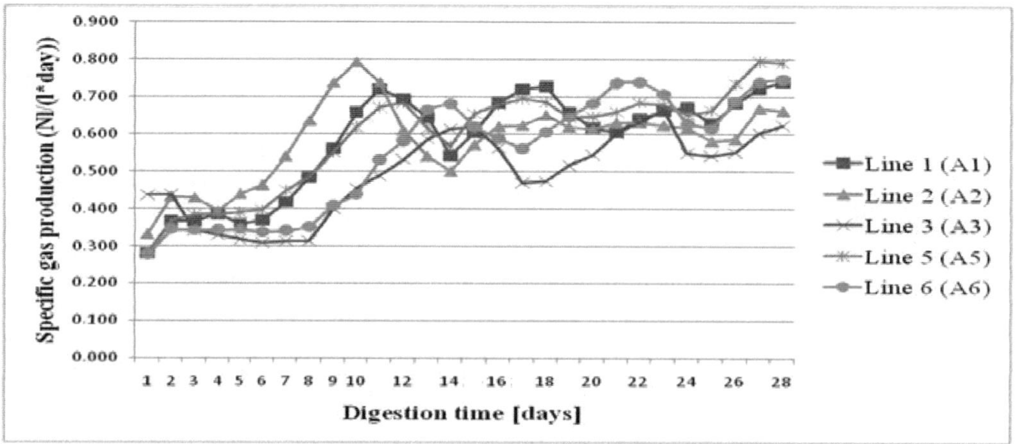


**Figure 4.** The variation of the specific gas production

Figure 4 presents the biogas quantity that was generated in the 28 days of the experiment. Although during the whole experiment all the 5 digesters were supplied with the same quantity of the material (66.1 grams of cow manure and 80.0 grams of water), the gas production varied. This fact is proved by the values recorded in figure 4, where one can observe the gas flow variations. According to the graphic the largest generated gas quantity was recorded at line 2 (A2) in the $10^{th}$ day, while the most reduced quantity was in the first day of the experiment at line 1 (A1).

## 3. Conclusions

The aim of this study was to determine the biogas quantity, which can be obtained from the cow manure. For this experiment there was used a laboratory installation consisting of 6 parallel digesters (A1-A6). The experiment was carried out for a period of 28 days, when the digesters were daily supplied. Throughout the whole experiment the pH was monitored, thus it could be supervised if the bacteria have an optimal environment of development.

During the experiment the recorded temperature was between 29.5 and 31.7°C, this fact indicates that the working environment, in which the experiment was carried out, was mesofilique. Although during the whole time of the experiment, the digesters were supplied with the same quantity of raw material, the obtain results show variations, concerning the quantity of the biogas. According to the experimental results, the most significant biogas quantity recorded was at line 2 (A2) in the $10^{th}$ day.

The result obtained in the achieved study on the cow manure proves the fact that it represents an important energy carrier and pillar and a real possibility concerning the biogas production.

## 4. References

[1] Behera S S and Ray R C, 2021 Bioprospecting of cow dung microflora for sustainable agricultural, biotechnological and environmental applications*CRMICR***2** 100018

[2] El-Mashad H M, van Loon W K, Zeeman G and Bot G P A 2005 Rheological properties of dairy cattle manure *Bioresour. Technol.***96** (5) 531-5

[3] Tan J B, Jamali N S, Tan W E, Che Man H and Abidin Z Z 2021 Techno-economic assessment of on-farm anaerobic digestion system using attached-biofilm reactor in the dairy industry *Sustainability* **13** (4) 2063

[4] Bhatt K and Maheshwari D K 2019 Decoding multifarious role of cow dung bacteria in mobilization of zinc fractions along with growth promotion of C. annuum L.*Sci. Rep.* **9** (1) 14232

[5] Maj I, Kalisz S, Szymajda A, Łaska G and Gołombek K 2021 The influence of cow dung and mixed straw ashes on steel corrosion *Renew. Energ.***177** pp 1198-1211

[6] Pochwatka P, Kowalczyk-Juśko A, Sołowiej P, Wawrzniak A and Dach J 2020 Biogas plant exploitation in a middle-sized dairy farm in Poland: Energetic and economic aspects *Energies* **13** (22) 6058

[7] Baek G, Kim D, Kim J and Kim H, Lee C 2020 Treatment of cattle manure by anaerobic co-digestion with food waste and pig manure: Methane yield and synergistic effect *Int. J. Environ. Res. Public Health* **17** (13) 4737

[8] Li L, Xu J, Wang H, Liu X and Zhang Di 2019 Study of the performance of biogas production by mixed fermentation of cow dung, deer manure, and mushroom fungus *Energy Sci. Eng.* **8** pp 466-475

[9] Kapoor R, Ghosh P, Tyagi B, Vijay V K, Vijay V, Thakur I S, Kamyab H, Nguyen D D and Nadda A K 2020 Advances in biogas valorisation and utilization systems: A comprehensive review *J. Clean. Prod.* **273** 123052

[10] Bote M A, Naik V R and Jagdeeshgouda K B 2020 Production of biogas with aquatic weed water hyacinth and development of briquette making machine *Mater. Sci. Technol.***3** pp 64-71

[11] Palve A M, Arukula R and Gupta R K 2021 Sustainable Bioconversion of Waste to Value Added Products, *Bioconversion of Biowastes for Energy Application. In: Inamuddin*, ed Inamuddin and A Khan, Springer (Switzerland) pp 1-22

[12] Dong L, Cao G, Guo X, Liu T, Wu J and Ren N 2019 Efficient biogas production from cattle manure in a plug flow reactor: A large scale long term study *Bioresour. Technol.***278** pp450-455

[13] Ma M, Wang J, Song X, Su W, Bai Y, Yu G 2020 Co-gasification of cow manure and bituminous coal: A study on reactivity, synergistic effect, and char structure evolution *ACS Omega***5** (27) pp 16779-16788

[14] Işık E H B and Polat F 2018 Effects of pretreatments on the production of biogas from cow manure *Int. adv. res. eng. j.***02** (01) pp 48-52

[15] Kougias P and Angelidaki I 2018 Biogas and its opportunities- A review *Front. Environ. Sci. Eng.***12**

[16] Lee M E, Steiman M W and St. Angelo S K 2020 Biogas digestate as a renewable fertilizer: effects of digestate application on crop growth and nutrient composition *Renew. Agric. Food Syst.* 36pp 173-181

[17] Bashir S et al 2021 Comparative role of animal manure and vegetable waste induced compost for polluted soil restoration and maize growth *Saudi J. Biol. Sci.***28** pp 2534-2539

[18] Liebetrau J, O'Shea R, Wellisch M, Lyng K A, Bochmann G, McCabe B K, Harris P W, Lukehurst C, Kornatz P and Murphy J D 2021 Potential and utilization of manure to generate biogas in seven countries *TCP by IEA* ed IEA Bioenergy Task **37**

[19] Adekunle A S, Ibitoye S E, Omoniyi P O, Jilantikiri L J, Sam- ObuC V, Yahaya T, Mohammad B G, Olusegun H D 2019 Production and testing of biogas using cow dung, jatropha and iron filins *JB&B***4** (3) 143

[20] Kweinor Tetteh E and Rathilal S 2021 Application of biomagnetic nanoparticles for biostimulation of biogas production from wastewater treatment *Mater. Today*: *Proc.* **45** 5214-20

[21] Rani P, Pathak V V and Bansal M 2021 Co-digestion of what straw and animal manure pretreated with calcium hydroxide for biomethane production: Kinetic study *CRGSC***4** 100145

[22] Cruz I A, Santos Andrade L R, Bharagava R N, Nadda A K, Bailal M, Figueiredo R T and Romanholo Ferreira L F 2021 Valorization of cassava residues for biogas production in Brazil based on the circular economy: An updated and comprehensive review *Cleaner Engineering and Technology* **4** 100196

[23] Jomnonkhaow U, Uwineza C, Soufiani A M, Wainaina S, Reungsang A and Taherzadeh M J 2021 Membrane bioreactor-assisted volatile fatty acids production and in situ recovery from cow manure *Bioresour. Technol.***321**124456

[24] Operation and Maintenance Manual 2017 BioReactor Simulator

[25] Wainaina S, Awasthi M K, Horváth I S and Taherzadeh M J 2020 Anaerobic digestion of food waste to volatile fatty acids and hydrogen at high organic loading rates in immersed membrane bioreactors *Renew. Energy***152** 1140-48

[26] Zlateva P, Terziev A and Yordanov K 2021 Study of regime parameters of the fermenter in the production of biogas from animal liquid waste materials *E3S Web Conf.***286** 02010

**Acknowledgments**
This research was carried out within the doctoral program of the first author, at UPT.

IOP Conf. Series: Earth and Environmental Science **960** (2022) 012017 doi:10.1088/1755-1315/960/1/012017

# Challenges and performance evaluation of livestock waste energy content

I A Halmaciu[0000-0002-9932-7366], I Ionel[0000-0002-2999-1210], I Vetres[0000-0002-6425-9818], R M Balogh[0000-0002-3788-8889] and D Bisorca[0000-0001-8879-9774]

[1]Politehnica University of Timisoara, Romania, Fac. de Mecanica, Bv. M. Viteazu, Nr. 1, 300222, Timisoara, Romania

E-mail: ioana.ionel@upt.ro

**Abstract.** The global increase of the population has generated more and more requirement of the animal-based food. In order to provide this requirement, it was necessary to increase considerably the actual numbers of animals. This has led to both numerous positive and negative effects brought both to people and animals. Creating agro-touristic farms, ensuring fresh food, creating workplaces are just a part of the factors which have beneficial effects on the human beings. Yet, a major problem, which should not be ignored and neglected, is represented by the wastes resulted from animal breeding. These superficially treated wastes can cause numerous negative effects on the whole ecosystem. The animal dejections, the water resulted from the meat processing, the animal corpses, and all represent biodegradable wastes, which might be used, by transforming their energy content into electrical and thermal energy. This can be possible, for example, by using these wastes as raw material for producing biogas. To prove their efficiency in producing the biogases there have been done thermal analysis. In this article are analysed, from a thermal point of view, 3 different samples. In the first part of the experiment were analysed the swine dejections, in the second part were analysed the poultry dejections, and the third experiment consisted in the analysis of the cow stable waste. The analysis of these samples was done with the help of the Netzsch 449 C Jupiter device. The results obtained from the analysis prove the fact that all the three raw materials can be used as raw materials in producing the biogas.

## 1. Introduction

The continual development of the zootechnical exploitation can have a positive effect on the population and on the environment, but at the same time, can have negative effects, if the wastes produced by the farms are not managed properly.

The intensive breeding of the animals can have positive effects on the population, not only by assuring the fresh food at accessible prices, but also by creating workplaces, especially in rural areas. The main meat food resources of the world come from the farm animals such as: swine, cattle, and poultry, in small quantities, sheep, and goats [1].

More than that, the developing of agro-touristic farms constitutes a recreation and relaxation environment, these having a positive impact on the human health. The agro-touristic farms can also contribute to the developing of the communities by offering the farmers possibilities to sell their own products, this thing bringing them supplementary income [2].

Regarding the negative impact of the animal breeding has on the environment; the recent studies have proved the fact that the animal production generates 14.5% from the total anthropogenic emissions of gas with greenhouse effects. Half of these come directly from the animal production, and the other half comes from the fodder production [3].

These emissions can be found in form of carbon dioxide ($CO_2$), nitrous oxide ($N_2O$), methane ($CH_4$) and ammonia ($NH_3$) [4].

Cleaning and ventilating the market hall, removing of the stable waste [5] transferring the animals to the breeding, delivering the animals to slaughter houses, preparing and the transport of the fodders are just a small part of the polluting sources with emissions. The high concentration of particles in suspension (PM) can affect negatively the human health by producing disorders of the central nervous system, interstitial pulmonary diseases, nose symptoms, asthma or even the toxic syndrome with organic dust (ODTS) [6, 7].

The animal nourishment also plays an important role in diminishing the total nitrogen excreted, and thus the ammonia emissions. To reduce the total nitrogen excreted it is necessary the reduction of the raw proteins content by applying a food regime balanced in nitrogen [8].

In order to prevent the diseases and to stimulate the animal breeding a major part of the animal breeders go in for administrating, sometimes excessively, of antibiotics. The studies have proved the fact that most of the antibiotics administrated to the animals are eliminated from the body through urine and excrement and that only a small part of these prevent the diseases. The main antibiotics administrated to the animals are: ionophores, peptides, macrolides [9] and tetracycline [10].

Until recently the animal waste was spread directly on the soil in order to administrate nutrients necessary for the agricultural crops. For example, in the composition of the poultry waste comes the nitrogen (N), phosphor (P), potassium (K), calcium (Ca), chlorine (Cl), boron (B), all these elements ensure a good development of the crops [11]. Unfortunately, the recent studies have shown the fact that the excreted waste by animals contains a significant quantity of residual antibiotic, which applied directly on the soil, without a preliminary treatment can cause significant damage to the environment, and not only [12, 13].

The presence of the antibiotics in different layers of the soil is called `terra-accumulation`. This occurs when the antibiotic reaches the soil at a rate that exceeds its degradation rate [14].

Storing and managing improperly the animal corpses in the case of some contagious diseases such as: the African swine pest, the poultry pest or mad cow disease, can lead to the propagation of the pathogen germs through a number of ways, such as the attraction of the flies. The liquids also resulted from wrong disposal of the corpses can constitute a contamination source of the soil and the groundwater [15].

The zootechnical farms which improperly manage the way of generating and eliminating of the wastes, represent the main source of polluting the rivers and the lakes with nitrites ($NO_3$) and phosphate ($PO_4$) [16, 17].

Another major source of polluting the waters represents the wastes coming from the slaughter houses. The used waters resulted from the processing of the meat, infiltrated in water can become a real danger for the aquatic environment. The slaughter houses wastes are affected by the zoonotic diseases, these animal diseases can be transmitted to human beings [18].

Although the humankind needs water and food, and the farms ensure a major part of the necessities, the wastes resulted from the industrial process of food administrated improperly can have irreversible effects on the ecosystems.

Recently the greater and greater generating of the animal wastes has stimulated the requirement of potential application for their managing. The measure that could solve the elimination of the wastes in an environmentally-friendly way consists in the biogas production. By this method the wastes are not only eliminated, but they are also changed into real sources of producing the electrical and thermal energy.

Thus, the production of bio-fuels from zootechnical wastes was designed as a possibility to manage the waste and their usage in new recycling sources of energy [19]. In order to obtain bio-methane of high quality it is necessary for the bio-fuel to get through a process of eliminating the impurities [20].

As a result of the producing process of the biogas there results a secondary product named compost, which is rich in nutritive substances, and which is a good alleviate for the soil [21, 22]. Unlike the untreated wastes, the compost is less smelly, it contains a small number of pathogen agents and it reduces the chance of burning the plant [23].

Unlike other wastes, the wastes resulted from animals are biodegradable and ensure a wide variety of sources [24] more than that they can be endlessly used.

Figure 1 synthesises the impact that the main sources of pollution generated by farms is produced on the ecosystems.

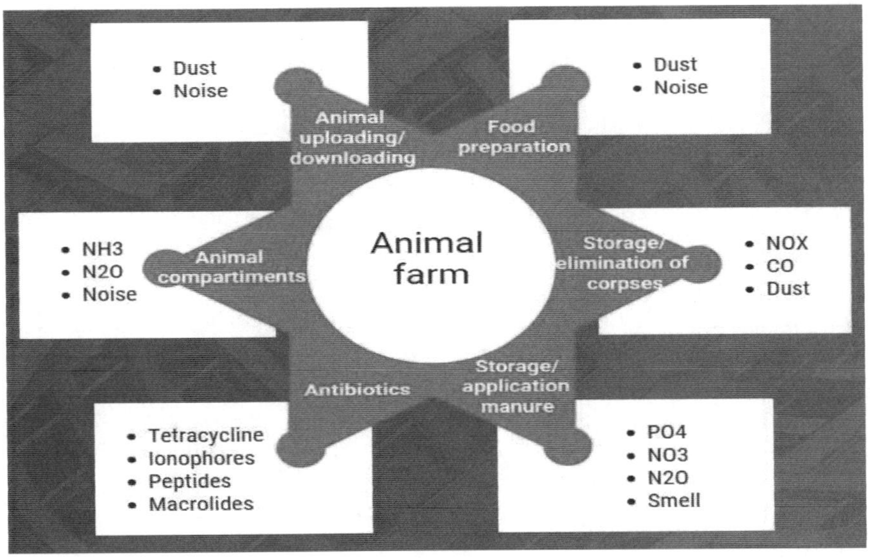


**Figure 1.** The scheme of the main pollutants generated by the farm.

## 2. Materials and methods

### *2.1. Materials*

During this experiment there were tested the animal wastes resulted from three different farms.

For the first experiment there were analysed the swine dejections resulted from a swine fattening farm.

During the second experiment there was analysed the poultry waste resulted from an intensive breeding poultry farm.

In the third experiment there was analysed the cow stable waste resulted from a mini family farm.

### *2.2. The description of the laboratory device*

The device Netzsch 449 C Jupiter has the capacity to record using a single sample for thermogravimetry analysis (TGA) and the calorimetry with differential scanning (DSC). The temperature interval that this device can cover is between -120° and 1650°C. The main function of the device consists in the fact that it continuously monitors the weight losses of the sample according to the temperature.

The thermogravimetry analysis is an alternative through which can be determined the energetic quality of the materials, using a small quantity of materials from 1 mg to several g, while the temperature in the oven increases up to1600°C [25, 26].

The calorimetry with differential scanning is a typical method of thermal analysis used for the direct assessment of the thermal energy absorption, which is produced between a test and a sample, during a controlled temperature [27, 28].

### *2.3. Describing the working process*

In order to achieve the thermogravimetry analysis (TGA) and the differential scanning calorimetry (DSC) for the three raw materials, it was necessary that a small quantity from each raw material (approximately 16 g), to be grinded to dust. The two analyses were carried out with the help of Netzsch STA 449 C Jupiter device, which offers the possibility to simultaneously record TG and DSC curves.

These curves were recorded in the air atmosphere, and the used crucibles were made of alumina. The temperature interval was between: 25- 1000°C, while the heating speed was 10°C/minute.

### *2.4. The derivatogram analysis*
Next there are presented the three derivatograms, as well as their interpretation.

#### *2.4.1. The thermal analysis of the swine dejection*

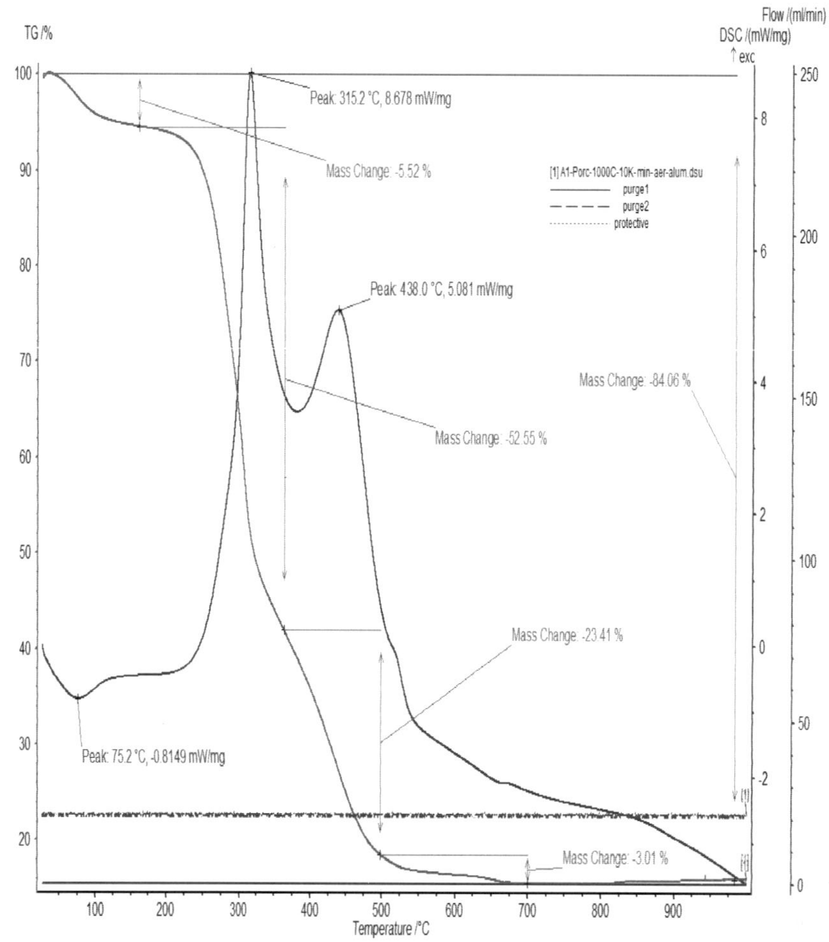


**Figure 2.** The derivatogram of the swine dejections sample.

The endothermic effect emphasized on the DSC curve at 75.2°C and accompanied by the mass loss of 5.52 % on the TG curve is attributed to the water evaporation (humidity). The DSC curve emphasizes two strong exothermic effects at 315.2°C and also at 438°C that can be attributed to the burning of the organic components in the sample.

The first exothermic effect is stronger, accompanied by a significant mass loss of 52.55% on the TG curves, and the second presents a mass loss of 23.41%. Between 500 and 700°C there can be also observed two small exothermic effects of a shoulder shape, the first being situated at approximately 510°C, and the second at 680°C, that can be attributed to the burning of the latest organic wastes.

These two effects are accompanied by a mass loss of 3.01%. Over 700°C there aren't emphasized the thermal effects, and the mass sample isn`t altered.

#### *2.4.2. The thermal analysis of the hen waste*
The endothermic effect emphasized on the DSC curve at 80.1°C and accompanied by a mass loss of 4.51% on the TG curve is attributed to the water evaporation (humidity) from the sample. The DSC curve is emphasized by a succession of exothermic effects that can be attributed to the burning of the

organic components in the sample. The first exothermic effect, situated at 307.5°C is accompanied on the TG curve by a mass loss of 35.75%.

At 428.5°C on the DSC curve there can be observed an exothermic shoulder followed by a strong exothermic effect situated at 464.0°C. The mass loss compatible with these effects is of 22.0%.

At a temperature of 704.3°C there can be observed an exothermic effect of small intensity, accompanied by a mass loss of 5.58%, attributed to the burning of the latest organic wastes.

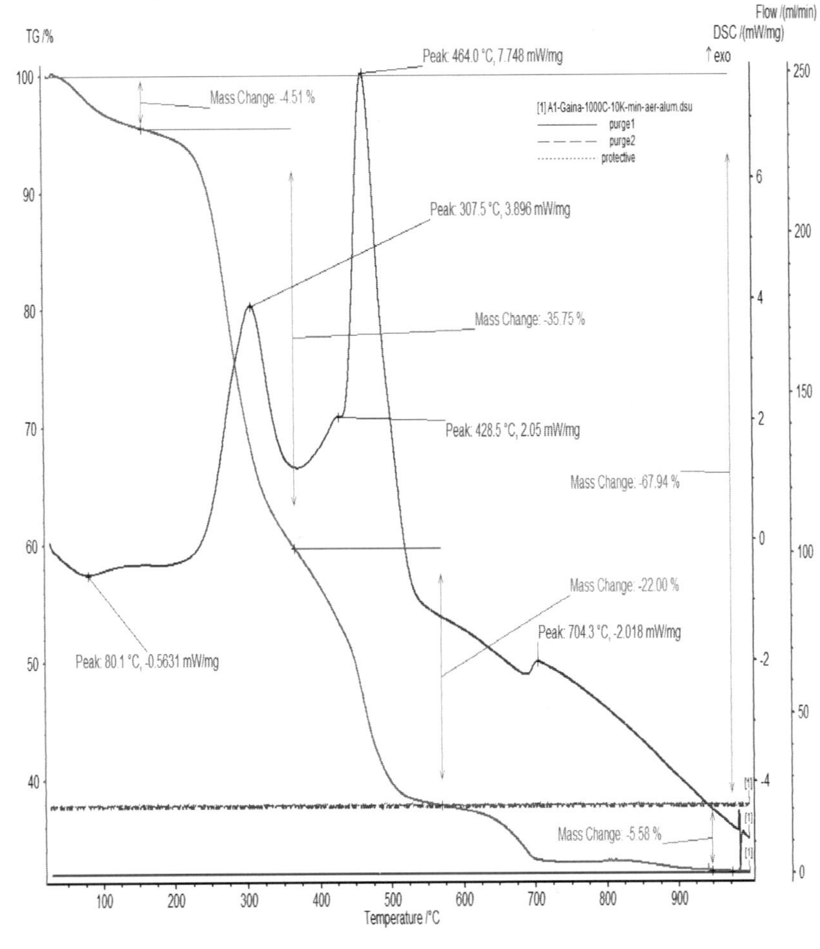


**Figure 3.** The derivatogram of the poultry waste sample [29].

### *2.4.3. The thermal analysis of the cow stable waste*

The endothermic effect emphasized on the DSC curve at 76.8°C and accompanied by the mass loss of 4.62% on the TG curve is attributed to the water evaporation in the sample. The DSC curve emphases three exothermic effects, that can be attributed to the burning of the organic components in the sample. The strongest exothermic effect is situated at 305.6°C and is accompanied on the TG curve by the greatest mass loss of 54.91%.

The next exothermic effect of a smaller intensity is situated at 369.8°C and is accompanied by a mass loss of 18.24%, and the least intense situated at 489.1°C is accompanied by the least mass loss of4.65%. Over 700°C there aren't emphasized thermal effects. To be noticed that on the bottom of the crucible there was stuck a small quantity of wastes and as well as 438°C that can be attributed to the organic components in the sample.

The samples analysed in Table 1 were not dried at room temperature.

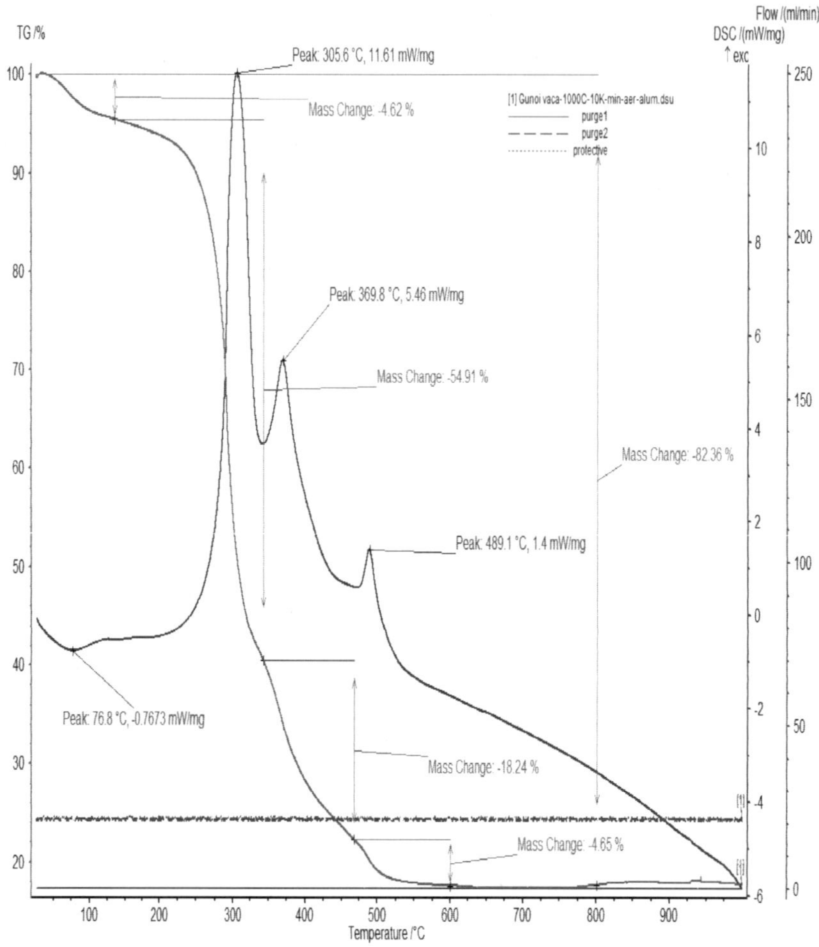


**Figure 4.** The derivatogram of the stable cow waste.

**Table 1.** Values resulting from derivatograms.

| The sample type | Humidity % | Contents of the volatile material % | Fixed carbon % |
|---|---|---|---|
| Swine dejections | 3 | 81 | 16 |
| Poultry waste | 3 | 60 | 37 |
| Stable cow waste | 4 | 78 | 18 |

## 3. Conclusions

The requirement on the meat products market made the farmers increase their animal goods, thing that led to a massive increase of the quantities of the zoo-technical wastes.

The animal wastes, the animal corpses and the evacuated water from the slaughter houses are the most representative wastes which contribute to the global heating by the increasing of the greenhouse effect gases.

For all these wastes there must be urgently found solutions to eliminate them in a way that will affect the environment less. The production of biogas is the measure that applies perfectly to eliminate and changes the wastes into cheap sources of electrical and thermal energy. The compost resulted from the anaerobe digestion is a good fertiliser, being rich in nourishing substances, improves the soil airing and kills the seeds of the weeds.

For this experiment there were analysed three samples of animal wastes: swine dejections, poultry wastes, stable cow wastes. These tests were analysed with the help of the Netzsch STA 449 C Jupiter device, which has the function to concomitantly record the TG and DSC analysis. During the whole experiment the tests were heated at different temperatures noticing the decreasing rate of the sample mass.

Looking at the three derivatograms one can notice the fact that the first mass losses, which represent the humidity, were recorded at the beginning of the experiment, when the endothermic effect occurred. During these endothermic effects the temperature for the three tests was between 75.2°C and 80.1°C. The mass losses recorded for each sample was the following: swine dejections: 5.52%, poultry wastes: 4.51% and the stable cow waste: 4.62%. These endothermic effects were followed by several exothermic effects of different intensities.

In the case of the swine dejections the first exothermal peak, which was more intense occurs at a temperature of 315.2°C being accompanied by a mass loss of 52.55%. Regarding the poultry waste and the stable cow waste, the strongest exothermal effects were recorded at close temperature, 307.5°C, respectively 305.6°C.

Making a classification and comparison of the productivity for these three tests, it will look like that: 1- stable cow waste, 2- swine dejections and 3- poultry waste. The results obtained after the laboratory analysis proved the fact that all three raw materials analysed are proper to be used as raw material in producing the biogas.

## 4. References

[1] Iglesias R, Muñoz R, Polanco M, Díaz I, Susmozas A, Moreno A D, Guirado M, Carreras N and Ballesteros M 2021 Biogas from Anaerobic Digestion as an Energy Vector: Current Upgrading Development *Energies* **14** 2742

[2] Varga M D 2017 Posibilitati de imbunatatire a producerii si a consumului de carne de pasare in Judetul Satu Mare (teza de doctorat) Rezumatul tezei.pdf (fmvt.ro) accessed: 17.08.2021

[3] ***Animal breeding reduces environmental impacts with one percent per year - WUR accessed: 14.08.2021

[4] Dopelt K, Radon P and Davidovitch N 2019 Environmental Effects of the Livestock Industry: The Relationship between Knowledge, Attitudes, and Behavior among Students in Israel *Int. J. Environ. Res. Public Health* **16** 1359

[5] Roman M, Roman K and Roman M 2021 Spatial Variation in Particulate Emission Resulting from Animal Farming in Poland *Agriculture* **11** 168

[6] Manisalidis I, Stavropoulou E, Stavropoulos A and Bezirtzoglou E 2020 Environmental and Health Impacts of Air Pollution: A review *Front. Public Health* **8**

[7] Lavaine E, Majerus P, Treich N 2020 Health, air pollution, and animal agriculture *Rev Agric Food Environ Stud* **101** pp 517-528

[8] ***CL2017D0302RO0000010.0001.3bi_cp 1..1 (europa.eu) accessed:18.08.2021

[9] Kuppusamy S, Kakarla D, Venkateswarlu K, Megharaj M, Yoon Y-E and Lee Y B 2018 Veterinary antibiotics (Vas) contamination as a global agro-ecological issue: A critical view *Agric. Ecosyst. Environ.* **257** pp 47-59

[10] Zalewska M, Blazejewska A, Czapko A and Popowska M 2021 Antibiotics and Antibiotic Resistance Genes in Animal Manure-Consequences of Its Application in Agriculture *Front. Microbiol.* **12** 610656

[11] Chastain J, Camberato J C and Skewes P 2021 3-b-1 Poultry Manure Production and Nutrient Content Nutrient content of Poultry Manure

[12] Tian M, He X, Feng Y, Wang W, Chen H, Gong M, Liu D, Clarke J L and Van Eerde A 2021 Pollution by Antibiotics and Antimicrobial Resistance in Livestock and Poultry Manure in China, and Countermeasures *Antibiotics* **10** 539

[13] Chen C, Pankow C A, Oh M, Heakth L S, Zhang L, Du P, Xia K and Pruden A 2019 Effect of antibiotic use and composting on antibiotic resistance gene abundance and resistome risks of soils receiving manure-derived amendments *Environ Int.* **128** pp 233-243

[14] Cristina T R, Doma O A and Mosneag C L 2012 Use of antibiotics and about quinolones in veterinary therapy *Vet. Drug* **6**

[15] Miller L and Flory G 2018 Administrator (management) of corpses in small and medium-sized livestock farms. Practical considerations *FAO* **13**, (original in Romanian: Gestionarea (managmentul) cadavrelor in fermele de cresterea animalelor de dimensiuni mici si medii. Consideratii practice)

[16] Arango-Osorio S, Vasco-Eceverri O, López-Jiménez G, González-Sanchez J and Isaac-Millán I 2019 Methodology for the design and economic assessment of anaerobic digestion plants to produce energy and biofertilizer from livestock waste *Sci Total Environ.***685** pp1169-1180

[17] "Ex-ante evaluation of NRDP 2014-2020" CPV Code 79414000-0 Strategic Environmental Assessment Environmental Report 2014 NRDP (original in Romanian: "Evaluarea ex-ante a PNDR 2014-2020" Cod CPV 79414000-4 Evaluarea Strategica de Mediu Raport de mediu 2014 *PNDR)*

[18] Orisakwe O E 2018 Nigeria: Environmental Health Concerns *Reference Module in Earth Systems and Environmental Sciences*

[19] Chowdhury T, Chowdhury H, Hossain N, Ahmed A, Hossen M S, Chowdhury P, Thirugnanasambandam M and Saidur R 2020 Latest advancements on livestock waste management and biogas production: Bangladesh's perspective *J. Clean. Prod.* **272** 122818

[20] Sumardino S, Adisukmo G, Hanif M, Budiyono B and Cahyono H 2021 Effect of Pretreatment and Ratio of Solid Sago Waste to Rumen on Biogas Production through Solid-State Anaerobic Digestion *Sustainability* **13** 7491

[21] Lee M E, Steiman M and St. Angelo S K 2021 Biogas digestate as a renewable fertilizer: Effects of digestate application on crop growth and nutrient composition *Renew. Agric. Food Syst.* **36** pp 173-181

[22] Logan M and Visvanathan C 2019 Management strategies for anaerobic digestate of organic fraction of municipal solid waste: Current status and future prospects *Waste Manag Res.* **37** pp 27-39

[23] Kaufman J, Roach A, Moreland J and Parcell J 2020 Report to: Missouri Agricultural and Small Business Development Authority Missouri Value added Grant Program *Biogas Digestion: Economic and Asset Assessment for Missouri*

[24] Ma M, Bai Y, Song X, Wang J, Su W, Yao M and Yu G 2020 Investigation into the co-pyrolysis behaviours of cow manure and coal blending by TG-MS *Sci. Total Environ.* **728** 138828

[25] Brand MA, Barnasky R R de S, Carvalho C A, Buss R, Waltrick D B and Jacinto R C 2018 Thermogravimetric analysis for characterization of the pellets produced with different forest and agricultural residues *Cienc. Rural* **48**

[26] Saadatkhah N, Garcia A C, Ackermann S, Leclerc P, Latifi M, Samih S, Patience G S and Chaouki J 2019 Experimental methods in chemical engineering: Thermogravimetric analysis- TGA *Can J Chem Eng.* **98** pp 34-43

[27] Chountoulesi M, Naziris N, Pippa N, Pispas S and Demetzos C 2019 Differential Scanning Calorimetry (DSC): An Invaluable Tool for the Thermal Evaluation of Advanced Chimeric Liposomal Drug Delivery Nanosystems *Thermodynamics and Biophsics of Biomedical Nanosystems (Applications and Practical Considerations)* ed C Demetzos and N Pippa Chapter 9 pp 297-337

[28] Gill P, Moghadam T T and Ranjbar B 2010 Differential scanning calorimetry techniques: applications in biology and nanoscience *J Biomol Tech* **21** pp 167-193

[29] Halmaciu I A, Ionel I, Wachter M R and Wachter A R 2021 Comparative analysis of the energy content of diverse fuels by using thermo-gravimetric investigations accepted for publication *J. Phys Conf. Ser.*

IOP Conf. Series: Earth and Environmental Science **960** (2022) 012017 doi:10.1088/1755-1315/960/1/012017

**Acknowledgments**
The authors thank for the support offered to Professor Dr. Eng. Pacurariu Cornelia-Silvia from the Faculty of Industrial Chemistry and Environmental Engineering within the Politehnica University of Timisoara.

# Water aeration researches using fine air bubble generators

**Nicolae Vlad SIMA, Nicolae BĂRAN, Nicoleta Dorina ALBU, Mihaela PETROȘEL and Mihaela CONSTANTIN***
University POLITEHNICA of Bucharest, Faculty of Mechanical and Mechatronics Engineering, Splaiul Independenței 313, sector 6, Bucharest, Romania

E-mail: i.mihaelaconstantin@gmail.com
ORCID: https://orcid.org/0000-0002-5685-3614

**Abstract.** The paper presents experimental researches able to demonstrate that the aeration installations with mobile fine bubble generators are more efficient than the classic ones with fixed, fine bubble generators. The paper develops water aeration researches by introducing two types of fine air bubble generators:
I. Fixed fine air bubble generators, which generate Ø <1 mm bubbles;
II. Mobile fine air bubble generators that move in the volume of water subject to aeration.
The theoretical and experimental results are presented for the two versions. According to experimental data, aeration with mobile FBG is more efficient in terms of oxygen transfer from atmospheric air to a stationary water mass.

## 1. Introduction

The water-to-air interfacial mass transfer can be performed using air bubbles that are dispersed in a volume of water using the air bubble generators.

The introduction of atmospheric air into water is a process necessary to ensure certain dissolved oxygen concentration in the water.

Oxygen dissolved in water is necessary for different life forms such as: bacteria, plants, fish, etc. which uses oxygen for breathing.

The oxygen dissolved in water necessary for life assurance is as follows [1]:
- fish need $4 \div 15$ mg / dm$^3$;
- oysters, crabs need $1 \div 6$ mg / dm$^3$.

The oxygen in the air is absorbed by the water in a slow process until it reaches saturation.

This process can be:
- naturally: waterfalls, waves, etc.;
- by forced means with installations produced on an industrial scale.

The aeration process can be carried out in two ways [1] [2] [3]:

A) By mechanical aeration.
B) By pneumatic aeration.

Mechanical aeration involves the operation of the aerators, with surface aeration.

Pneumatic aeration is more efficient and provides the introduction of air below the free water level; in this case, the air is introduced into the water layer by installations at the base of the water tank.

Pneumatic water aeration systems are divided into three classes [4] [5]:

I) Perforated pipes.
II) Porous diffusers made of ceramic materials, elastomeric membranes, etc.
III) Fine bubble generators (FBG) having as a dispersion element a flat plate provided with orifices.

The following will present two versions of class III:
I) Fine bubble generator, which is immobile (fixed) in a water tank.
II) Fine bubble generator, which is movable, i.e. it is moved by means of an electromechanical mechanism in a water tank.

The aim of the experimental researches is to demonstrate that the aeration installations with mobile fine bubble generators are more efficient than the classic ones with fixed, fine bubble generators. Aeration with mobile fine bubble generators is necessary especially where areas with low dissolved oxygen levels are formed in a large water tank (lakes, ponds, etc.), and an installation with fixed FBG would lead to additional operating costs.

Experimental researches are also being carried out to support the idea according to which an aeration system with fine, mobile bubble generators is more efficient than a fixed one, because air bubbles have a longer trajectory in the water mass. The contact between the air bubble and the water mass around it is practically increased, thus increasing the performance of the aeration system.

## 2. Bubble generator, which is immobile (fixed) in a water tank

For the two versions (I + II) the input data for the theoretical and experimental researches are the same:

- Air outlet section area in water: $A_0 = 1.2 \cdot 10^{-6}$ m$^2$;
- Compressed air flow rate in water: $\dot{V} = 0.6$ m$^3$/h;
- Hydrostatic load: H = 500 mm $H_2O$;
- Water and air temperature: t = 24.6 °C;
- Initial dissolved oxygen concentration in water: $C_0 = 5.46$ mg/dm$^3$;
- The duration of the oxygenation process, in accordance with the experimental research's methodology.

The constructive solution of the fine bubble generators (FBG) was made of transparent plexiglass, and their shape is similar to a parallelepiped provided with a plate with orifices and inlets for compressed air.

### *2.1. The constructive solution of FBG in version I*

Considering the previous researches [5] [6] a number of 17 orifices with $d_0 = 0.3$ mm was chosen, placed on a flat plate of s = 2 mm thickness.

The first necessary condition ($s/d_0 > 3$) for the FBG becomes [1]:

$$\frac{s}{d_0} = \frac{2}{0.3} = 7 \tag{1}$$

The second condition ($d/d_0 > 8$) for the FBG construction becomes:

The area of the outlet section ($A_0$) of the air in the water will be:

$$\frac{d}{d_0} = \frac{6}{0.3} = 20 \tag{2}$$

The initial diameter of the air bubble will be:

$$A_0 = n\frac{\pi \cdot d_0^2}{4} = 17\pi\frac{(0.0003)^2}{4} = 1.2 \cdot 10^{-6}[\text{m}^2] \tag{3}$$

where:
$D_0$ is the initial diameter of the air bubble [m];
$\sigma$ - surface tension [N/m];
$d_0$ - orifices diameter [m];
$\rho l$ - water density [kg/m$^3$];
$\rho g$ - air density [kg/m$^3$];
g - gravitational acceleration [m/s$^2$].
Since $D_0 < 1$ mm, it means that it will work with fine bubbles [1].

IOP Conf. Series: Earth and Environmental Science **960** (2022) 012018 doi:10.1088/1755-1315/960/1/012018

For the plate with orifices a rectangular shape was chosen (figure 1).

The following examples show how to format a number of different figure/caption combinations.

**Note that the table borders are shown as broken lines for guidance only.**

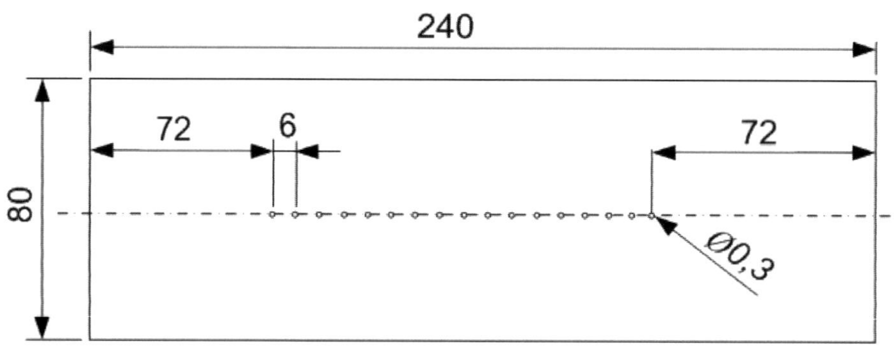


**Figure 1. Orifice plate.**

For this plate, a constructive form of the fine bubble generator was designed (Figure 2).

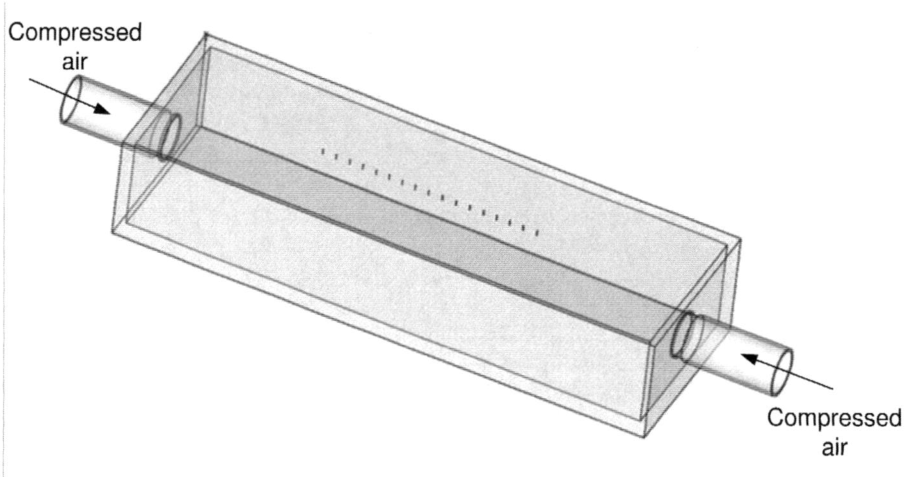


**Figure 2.** Overview of the FBG
1 - FBG body; 2 - plate with orifices; 3 - FBG connections for compressed air supply.

The fine bubble generator in figure 2 is mounted on the basin of the water basin, above which is a layer of water with the height H = 500 mm.

*2.2. Scheme of the experimental installation*
The scheme of the installation is shown in figure 3 and consists of:
-An electro compressor that compresses the air and sends it to the compressor tank.
-A pressure reducer.
-Compressed air plastic pipes for the air delivery, with inner diameter Ø = 15mm and a wall thickness of 2 mm; they supply the microbubble generator with air and ensure the evacuation of the surplus air delivered by the compressor to the atmosphere.
-Aeration tank made of plexiglass plates with a thickness of 5 mm, having the size of $0.5 \cdot 0.5 \cdot 1.6$ m.

-The fine bubble generator was subjected to tests in an experimental installation built in the laboratories of the University POLITEHNICA of Bucharest.

The orifices in the plate were made using a C.N.C. (Computer Numerical Center) which has a special machine KERN type for micro processing.

-The mechanism of action of the oxygenometer probe.

-Table with control electronics.

-Devices for measuring the flow rate, pressure, temperature and the dissolved oxygen concentration in water.

This concentration is measured every 15 minutes for two hours; from previous experiments in two hours or less from $(CO_2)_0$ reaches $C_s$.

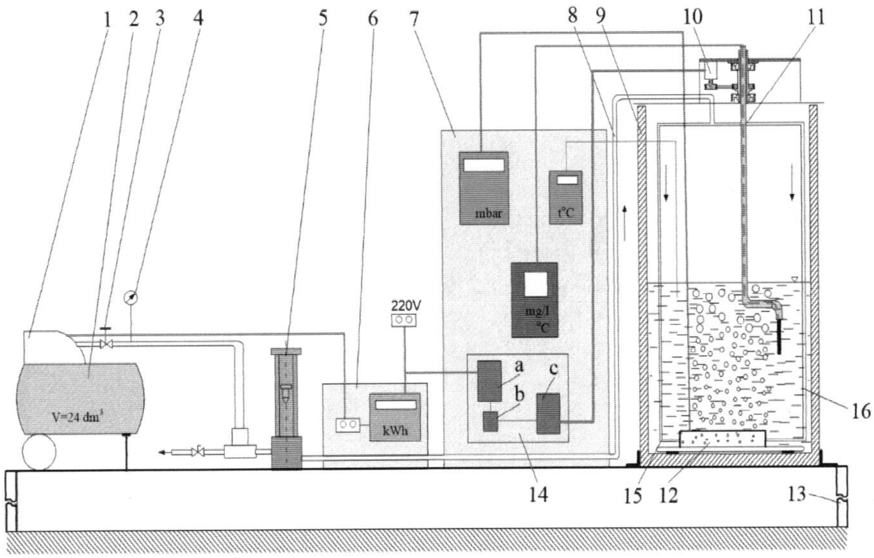


**Figure 3.** Scheme of the experimental installation regarding the water oxygenation
1 - electric compressor; 2 - compressed air tank; 3 - pressure reducer;
4 - manometer, 5 - rotameter; 6 - electric panel; 7 - panel with measuring devices; 8 - pipeline for the transport of compressed air to FBG; 9 - water tank; 10 - mechanism of actuation of the probe;
11 - oxygen meter probe; 12 - FBG of rectangular shape; 13 - support for the installation;
14 - control electronics: a - power supply, b - switch, c - control element; 15,
16 - FBG supply pipes with compressed air.

The air delivered by the electric compressor (1) passes through the pressure reducer (3) then through the rotameter (5) and through the pipes (15), (16) enters the water tank (9) and reaches the fine bubble generator (FBG) (12).

*2.3. The results of theoretical and experimental researches*

The differential equation of the rate of increase of the dissolved oxygen concentration in water has the form [6] [7] [8]:

$$\frac{dC}{d\tau} = ak_L(C_s - C)\left[\frac{kg}{m^3}\cdot\frac{1}{s}\right] \qquad (4)$$

where:

$aK_L$ is the volumetric mass transfer coefficient, $[s^{-1}]$;

$C_s$ - the mass concentration of oxygen at saturation in the liquid phase, $[kg/m^3]$;

C - current mass concentration of oxygen in the liquid phase, $[kg/m^3]$.

The relation (4) has been numerically integrated; a calculation program [9] was performed, the results being presented in figure 4, curve 1.

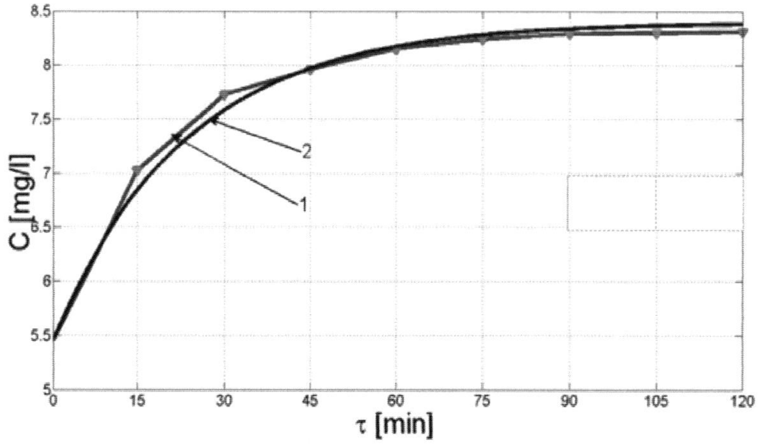


**Figure 4.** Increased of dissolved O2 concentration in water as a function of time
1 - graph constructed with data resulting from the calculation program;
2 - graph built with experimental data.

From Figure 4 a good coincidence is observed between the data resulting from the calculation and the data experimentally established.

## 3. Bubble generator that is mobile in a water tank

### *3.1. Initial dates*

When the FBG moves, the initial velocity of the air bubble entering the water will be (figure 5):

$$w_0 = w + w_{FBG} = \frac{\dot{V}_0}{\pi \cdot n \cdot r_0^2} + w_{FBG} \quad (5)$$

where:
w - is the air velocity when exiting the orifice;
$w_{FBG}$ - FBG movement speed;
n - the number of orifices;
$r_0$ - orifice radius.

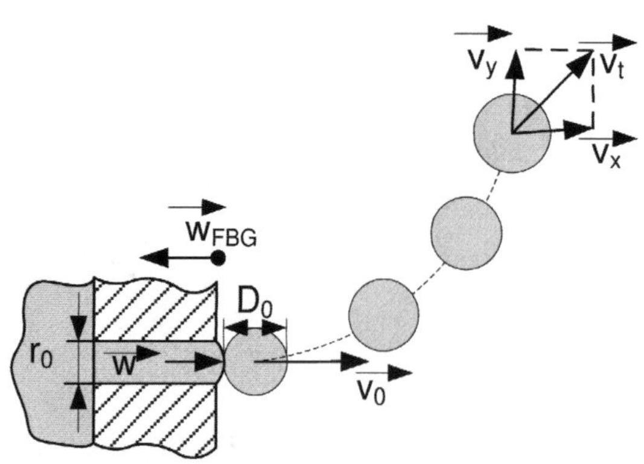


**Figure 5.** Initial speed of the air bubble for the mobile FBG.

In the tank where the fixed FBG is introduced, the water has a volume V = 0.5 · 0.5 · 0.5 = 0.125 $m^3$ (figure 6, a); this tank is oxygenated with a FBG for two hours; air flow is ($\dot{V}$ = 600 $dm^3/h$) and H = 500 $mmH_2O$.

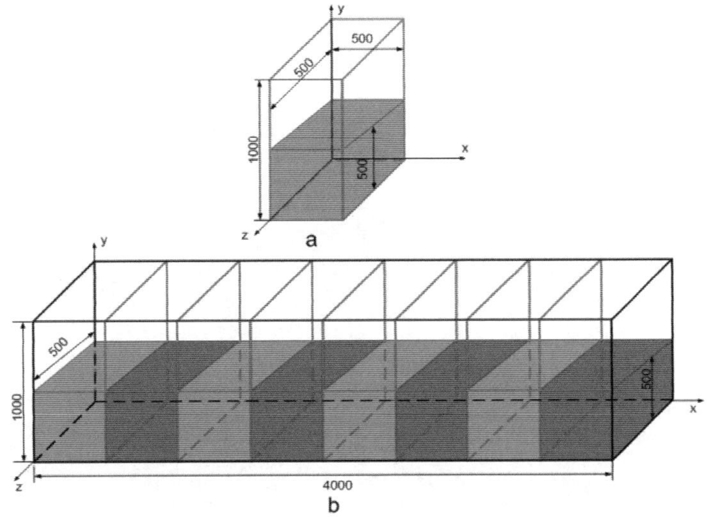


**Figure 6.** Sketch of the aeration tanks
a) aeration tank for immobile FBG; b) aeration tank for mobile FBG.

In this case, the $O_2$ concentration in water increases from $C_0$ = 5.46 mg/$dm^3$ to $C_{120}$ = 8.31 mg/$dm^3$ during two hours. If the tank is increased by the ox axis eight times one can obtain (figure 6, b) a volume of V = 8·0.125= 1 $m^3$.

If there is only one fixed FBG in this tank, to get from $C_0$ to $C_{120}$ a longer time is needed, theoretically equal to 8 · 2h = 16 h.

What happens if a mobile FBG with the same characteristics is introduced into the large tank and is aerated using the same flow rate as that used for fixed FBG?

From the experimental researches it was found that to reach from $C_0$ to the same value $C_{120}$ it does not take 16 hours, but only four hours. So one of the advantages of mobile FBG is that it reduces oxygenation time four times. In addition, the use of mobile FBG increases the degree of water turbulence leading to an increase in the rate of water oxygenation.

### *3.2. Experimental installation*

Figure 7 shows a general sketch of the experimental installation for mobile FBG researches.

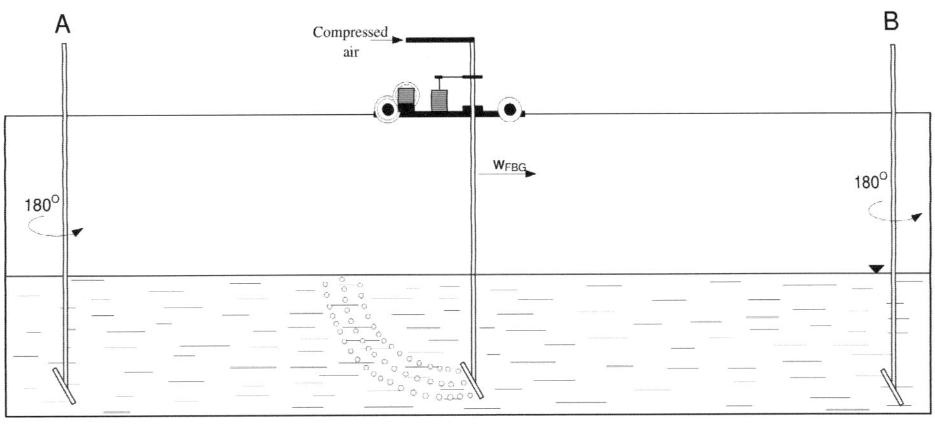


**Figure 7.** Experimental installation

1 - water tank; 2 - mobile platform; 3 - FBG compressed air supply line; 4 - mobile FBG;
A, B - the place where the FBG rotates 180°.

Figure 8 shows the mobile FBG in operation.

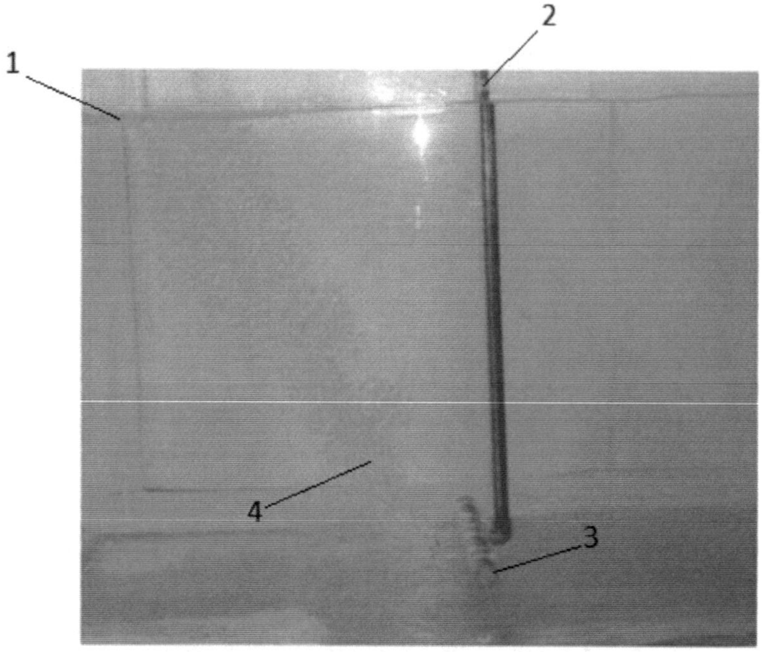


**Figure 8.** FBG in operation at $w_{FBG} = 0.2$ m/s
1 - the water level in the tank; 2 - air supply pipe; 3 - mobile FBG; 4 - the plane jet with air bubbles.

The measured oxygen concentration values are presented in table 1.

**Table 1.** O2 concentration values measured at different FBG movement speeds.

| τ [min] | | 0 | 30 | 60 | 90 | 120 | 150 | 180 | 210 | 240 | 270 |
|---|---|---|---|---|---|---|---|---|---|---|---|
| $C_{med}$ | $w_{FBG} = 0.2$ m/s | 5.47 | 6.55 | 7.16 | 7.54 | 7.83 | 8.04 | 8.21 | 8.32 | 8.36 | 8.40 |
| | $w_{FBG} = 0.4$ m/s | 5.45 | 6.23 | 6.84 | 7.22 | 7.55 | 7.82 | 8.13 | 8.23 | 8.30 | 8.36 |
| | $w_{FBG} = 0.6$ m/s | 5.45 | 6.19 | 6.79 | 7.19 | 7.51 | 7.81 | 8.02 | 8.18 | 8.25 | 8.31 |

where:
τ - is the time in [min]
$C_{med}$ – the measured dissolved oxygen concentration in [kg/m$^3$]

Based on the data in table 1, the graphs in figure 9 were constructed.

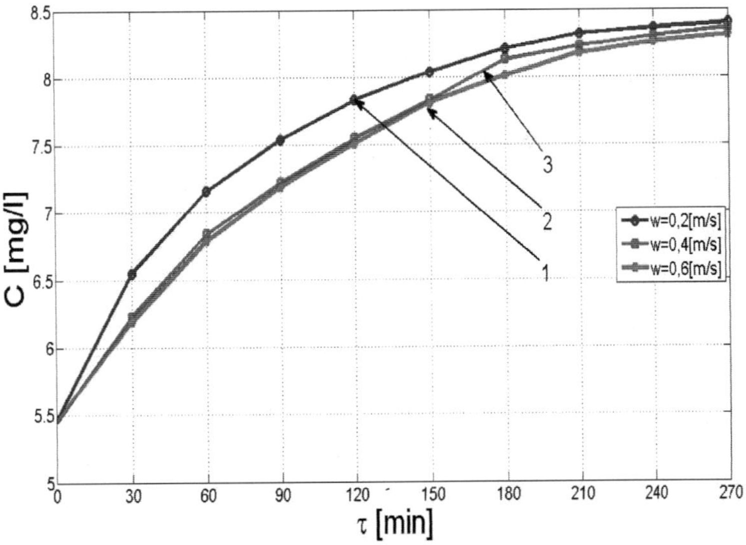


**Figure 9.** Variation of dissolved $O_2$ concentration in water in time, for different FBG speeds $w_{FBG} = 0.2$ [m/s]; $w_{FBG} = 0.4$ [m/s]; $w_{FBG} = 0.6$ [m/s].

Figure 10 compares the experimental results obtained for immobile and mobile FBG, obviously starting from the same initial data ($\dot{V}$, H, $C_0$, etc.). The data obtained are comparable to those in the literature [8] [10].

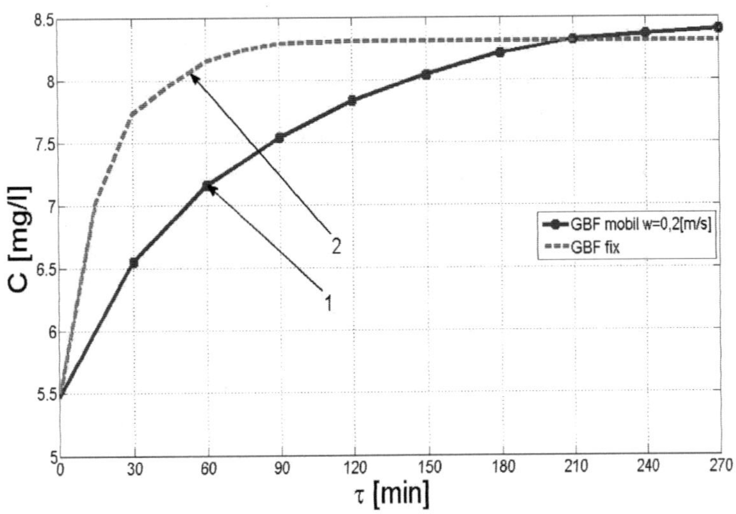


**Figure 10.** Variation of dissolved $O_2$ concentration in water in time, for different FBG speeds $w_{FBG} = 0.2$ [m/s]; $w_{FBG} = 0.4$ [m/s]; $w_{FBG} = 0.6$ [m/s].

So as can be seen in figure 10, aeration with mobile fine bubble generators is more efficient in the sense that the air bubbles transfer more oxygen to the water.

If for a volume of water of 0.125 $m^3$, in the case of fixed FBG, a value of 8.31 mg/ $dm^3$ of dissolved $O_2$ is reached in 120 min, in aeration with mobile FBG this value was reached after only 210 min at a volume of water eight times higher (1 $m^3$). A value of 8.40 mg/$dm^3$ of dissolved $O_2$ is reached in 240 min and not during 960 min as it would have resulted according to the similarity constant. So the oxygenation time is reduced four times.

## 4. Conclusions

1. Using FBG has the following advantages:
- It ensures a uniform distribution of air bubbles in the volume of water subject to aeration;
- Air bubbles are the same size;
- The diameter of the air bubbles can be controlled by the size of the outlet through the perforated plate;
- The pressure losses at the air passage through the FBG are much lower than the porous diffusers made of porous materials [1];
2. With the decrease of the outlet diameter of the air bubbles in the water, the rate of oxygen transfer to the water will increase [1].
3. Compared to fixed FBGs for the same initial data by using mobile fine bubble generators, the time in which the dissolved oxygen concentration in water reaches $C0 \rightarrow Cs$ is reduced four times.
4. According to experimental researches, the best aeration speed is 0.2 m / s, because, starting with 0.4 m / s and especially at 0.8 m / s, the coalescence of bubbles appears and the aeration process becomes more inefficient. However, even so, it is much more efficient than an aeration system with fixed FBG.
5. After aeration with mobile FBG, it was observed that the air bubbles for $w_{FBG} = 0.2$ m / s are evenly distributed without their coalescence, which intensifies the oxygenation.
6. By moving the FBG in the tank, the degree of turbulence increases, which leads to an increase in the transfer rate of dissolved oxygen to the water.

## 5. References

[1] Oprina G, Pincovschi I and Băran G 2009 *Hidro-Gas-Dynamics Aeration systems equipped with bubble generators*, POLITEHNICA PRESS Publishing House, Bucharest
[2] Oprina G 2007 *Contributions to the hydro-gas-dynamics of porous diffusers*, PhD Thesis, University POLITEHNICA of Bucharest, Faculty of Power Energy
[3] Robescu D, Robescu D and Verestoy A 2002 *Reliability of processes, installations and equipment's for water treatment and purification*, Technical Publishing House, Bucharest
[4] Rasha C 2017 *The influence of the architecture of fine bubble generators on the increase of the concentration of dissolved oxygen in water*, PhD Thesis, University POLITEHNICA of Bucharest
[5] Călușaru I M 2014 *The influence of the physical properties of the liquid on the efficiency of the oxygenation processes*, PhD Thesis, University POLITEHNICA of Bucharest
[6] Pătulea Al S 2012 *The influence of the functional parameters and of the architecture of the fine bubble generators on the efficiency of the aeration installations*, PhD Thesis, University POLITEHNICA of Bucharest
[7] Hand D W, Hokanson D R, Crittenden J C 2012 *MWH's Water Treatment: Principles and Design*, Third Edition, chapter 4- Air stripping and aeration, John Wiley & Sons, Inc
[8] Stenstrom M K, Leu S Y B and Jiang P 2006 Theory to Practice: Oxygen Transfer and the New ASCE Standard, *Proceedings of the Water Environment Federation* **15** 4838-52
[9] Pătulea Al, Călușaru I M and Băran N 2012 Researches regarding the measurements of the dissolved concentration in water, *Adv Mater Res* **550-553** 3388-94
[10] Băran N, Pătulea Al and Călușaru I M 2011 The Determination of The Oxygen Transfer Speed in Water in Nonstationary Conditions, International proceedings of Computer Science and Information Technology, Mechanical Engineering, Robotics and Aerospace 267-72

IOP Conf. Series: Earth and Environmental Science **960** (2022) 012019 doi:10.1088/1755-1315/960/1/012019

# The use of microtechnology's in the construction of water aeration installations

**Marilena Monica BOLTINESCU (Roza), Nicolae BĂRAN, Albertino Giovani ROZA and Mihaela CONSTANTIN***
University POLITEHNICA of Bucharest, Faculty of Mechanical and Mechatronics Engineering, Splaiul Independenței 313, sector 6, Bucharest, Romania

E-mail: i.mihaelaconstantin@gmail.com
ORCID: https://orcid.org/0000-0002-5685-3614

**Abstract**. Water aeration systems are highly efficient if the dispersion of air in the water is carried out in a controlled and uniform manner. The use of fine bubble generators ensures this and in addition, creates a small loss of pressure when air passes through them. The paper demonstrates that producing as few air bubbles as possible leads to a more efficient aeration process. Two water aeration installations are compared:
- The first has a perforated plate with 152 orifices Ø 0.1 mm;
- The second has four perforated plates, each with 113 orifices Ø 0.05 mm;
Both installations are successively supplied with the same flow rate of compressed air, at the same temperature and at the same initial dissolved oxygen concentration in the water.

## 1. Introduction

The need of water oxygenation is given by one of the problems faced by various stationary water sources, namely, eutrophication, a critical global environmental problem [1][2][3][4], which plays a major role in reducing the oxygen requirements of aquatic life. Green-blue algae are also called cyano-bacteria. It can produce their own nutrients through the photosynthesis process or they can use various nutrients from the environment. It can use phosphorus, nitrogen or potassium as a food source. It grows rapidly in late summer or early autumn, when it is very hot and produces *water eutrophication*. When blue algae's die, age or lack oxygen (they produce during the day, but consume at night), it quickly rots, produce gas, smell gas, swampy appearance, and *water oxygenation* is mandatory.
In water aeration systems the essential element is the water aeration device.

Water aeration can be achieved in three ways [5][6] namely:

a) by mechanical aeration;
b) by pneumatic aeration;
c) by mixed aeration.

Pneumatic aeration is more efficient and consists of introducing pressurized air bubbles into the aeration tank. According to the way the air bubbles are obtained, the aeration installations are classified as follows [7]:

A. Aeration installations with porous diffusers made of ceramic elements or elastic membranes;
B. Aeration installations made of pipes provided with orifices; these pipes are mounted on the radiator of the water tank;
C. Aeration installations with fine bubble generators.

**A. Aeration systems with porous diffusers.**

Fine bubble diffusers are immersed in water and ensure the dispersion of air in the form of bubbles. They are placed so that the entire body of water is agitated and mixed, as well as to provide the necessary oxygen required by the biological process.

Porous diffusers can be mounted evenly distributed over the entire surface of the on the radiator of the water tank, the air released through the diffusers rising to the surface of the water, producing turbulence that promotes gas transfer, or near the tank wall, the air inducing spiral currents in the water, which have the role of an agitator and improve the transfer. Porous diffusers can have different shapes: plates, discs, dome, tubes, etc. For their construction can be used special concrete, ceramic materials, plastics, porous metals, etc. Plate diffusers - are usually porous plates in the form of discs with a diameter of 300 mm and a thickness of 25 ÷ 38 mm. The plates are installed by grouting in the recesses in the floor, cementing or fastening in metal supports. The surface occupied by the porous plates is 15 ÷ 25% of the total radiator.

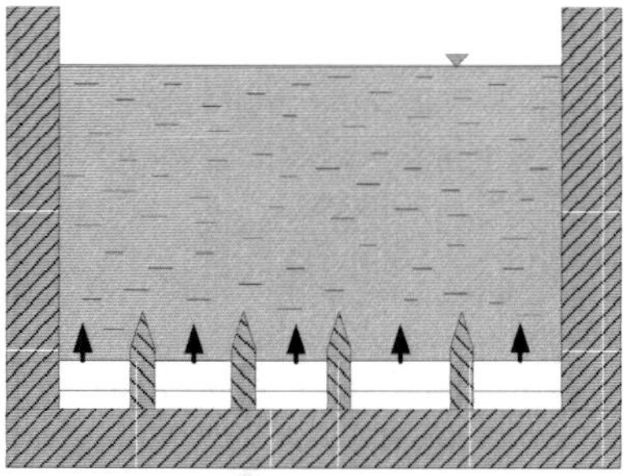

**Figure 1**. Porous plate aeration tank [7].

Rectangular plate diffusers are less used because it has:
- an uneven distribution of air with numerous plates attached at the same level;
- the inconvenience of removing the plate when it is grouted;
- the difficulty of adding other diffusers to meet future biomass increases.

Disc diffusers are available in diameters ranging from approximately 180 ÷ 240 mm and have a thickness of 13 ÷ 19 mm. The discs consist of two parallel flat surfaces. Uneven profiles help produce uniform airflow over the entire surface of the disc.

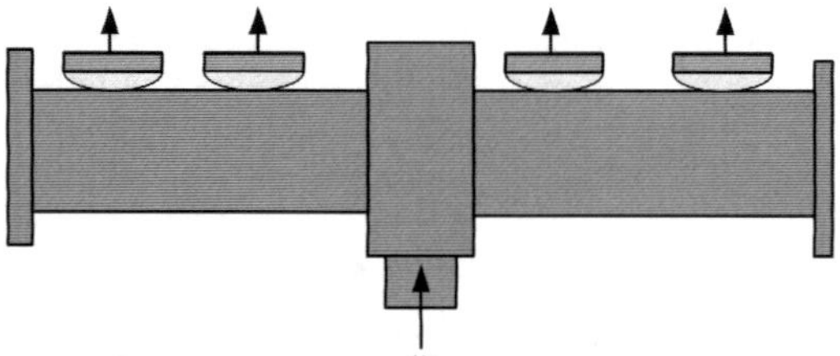


**Figure 2**. Disc porous diffusers [7].

The ceramic dome diffuser - is essentially a sectioned circular disc (figure 3). These diffusers are 180 mm in diameter and 38 mm high. It is mounted on PVC base plates that are welded to the air distribution pipes. In order to distribute the air better throughout the system, control orifices are placed in each diffuser assembly to create additional pressure loss and to balance the airflow. The domes are installed in a total floor covering or grid pattern. When oxygen demand is low, the diffusers can be placed in well-spaced rows along the side or in the middle of the tank, in a spiral. The diffusers are mounted as close as possible to the tank floor, usually 230 mm from the highest point on the floor [7].

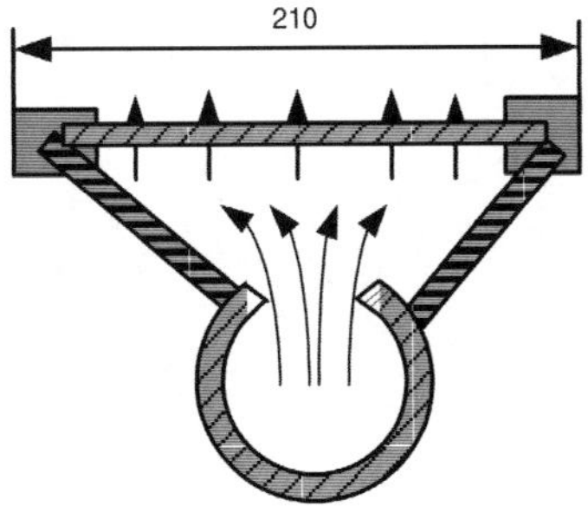


**Figure 3**. Dome-type porous diffusers [7].

Tubular type diffusers (figure 4) - are mounted at approximately 500 ÷ 700 mm from the tank radiator, in groups of 8 on a tilting arm around a joint. These diffusers have a length of 510 ÷ 610 mm and a diameter of 64 ÷ 76 mm. The component parts are made of corrosion-resistant plastic [7].

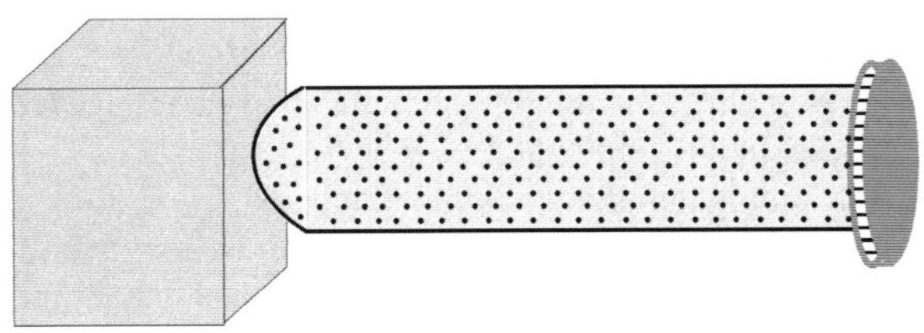

**Figure 4**. Tubular diffuser [7].

Porous diffusers must meet the following conditions:
a. high mechanical strength;
b. resistance to water action;
c. light mechanical processing;
d. high and uniform air permeability.

Choosing bubble size, diffuser type, and diffuser location can have a big effect on the efficiency of the aeration process.

**B. Aeration systems with pipes provided with orifices**

The pipes provided with orifices are placed perpendicular to the walls of the tank, at a certain distance from the water level; it is usually located on the tank radiator. A series of pipes are placed in the middle of the tank and along it, which helps to achieve a good mixture between air and water in the tank. The orifices of the pipes are of the order of 1 ÷ 3 mm.

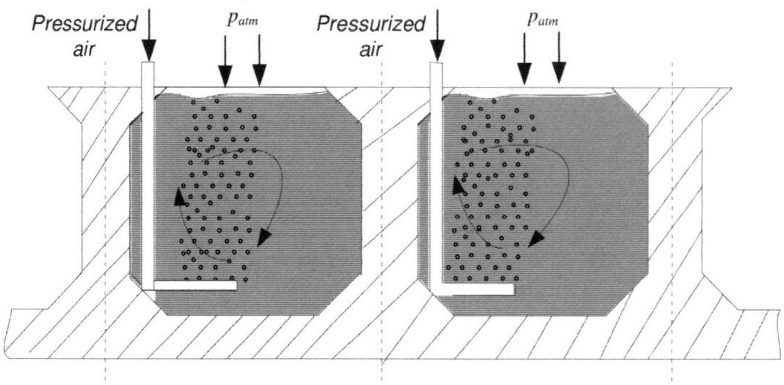


**Figure 5**. Pneumatic aeration tank with orifice pipes.

**C. Aeration systems with fine bubble generators**

Fine bubbles are bubbles created by orifices with Ø <1 mm. Currently, fine bubble generators (FBG) are made in the form of glass, ceramic, etc. diffusers. The diffusers have the disadvantages:

- does not ensure the supply of bubbles of equal diameter;
- does not ensure the supply of fine bubbles on the entire surface of the diffuser;
- the pressure losses when the air passes through them are large [7].

To eliminate these disadvantages, a team of researchers from the Department of Thermotechnics, Engines, Heating and Refrigeration Equipment researched [7]:

- the construction of fine bubble generators by using a laser;
- the construction of fine bubble generators by Ethylene Propylene Diene Monomer (EPDM);
- the construction of fine bubble generators with special working machines.

Thus, the following types of fine bubble generators were performed:

a) fine bubbles generator with 0.5 mm orifices, the plate with orifices being made of aluminum;

b) fine bubbles generator with 0.3 mm orifices, the plate with orifices being made of transparent plexiglass;

c) fine bubbles generator with 0.1 mm orifices, the plate with orifices being made of aluminum.

## 2. Formulation of the problem

Previously, in the researches carried out for the aeration of the waters, a fine bubbles generator, which had a plate with orifices of Ø = 0.1 mm, was used.

From the following it follows that the increase in the dissolved oxygen concentration in water is larger if the diameter of the air bubble immersed in water is smaller.

The rate of oxygen transfer to water is given by the relation [8], [9]:

$$\frac{dC}{d\tau} = a \cdot k_L \cdot (C_s - C) \text{ [kg/m}^3\text{s]} \quad (1)$$

relation where:

$$a = \frac{A}{V} [\text{m}^2/\text{m}^3] \quad (2)$$

A - the area of gas bubbles [$m^2$].

V - the volume of the biphasic system (air + water) [$m^3$].

$C_S$ - mass oxygen concentration at saturation [kg / $m^3$];

C - current mass oxygen concentration of in water at time τ [kg / $m^3$].

$k_L$ – the mass transfer coefficient [m / s].

Relation (1) indicates the change in oxygen concentration over time, due to the molecular diffusion of $O_2$ from the high concentration area to the low concentration area.

From relation (1) one can observe that in order to increase the transfer rate of O2 to water it is necessary:

I. the magnification of a, $k_L$ and $C_S$

II. the decrease of $C_0$

Conditions I and II are set out in Table 1.

From table 1 it results that in order to increase the concentration of dissolved O2 in water, it is necessary to decrease the diameter of the FBG orifices, therefore, implicitly to decrease the diameter of the air bubbles immersed in water. As a result, researches on the use of microtechnology's in water aeration processes will be considered by creating aeration installations that produce air bubbles in the range Ø = 0.05 mm [10].

**Table 1**. Solutions for increasing it $\frac{dC}{d\tau}$.

| No. | The pursued goal | The theoretical solution | The practical solution |
|---|---|---|---|
| 1 | The increase of $C_s$ | Increasing the $O_2$ concentration in the air introduced into the water | Introduction of oxygen, ozone ($O_3$) into water |
| 2 | The decrease of $C_0$ | Minimum values for $C_0$ depending on the nature of the micro-organisms present in the water | Decreased initial water temperature; Introduction of $C_0$-reducing substances into water |
| 3 | The increase of $k_L$ | Intensification of turbulence | FBG rotation<br>Using mobile FBG |
| 4 | The increase of a | Decrease the gas bubble diameter | Decreasing of the FBG orifices diameter |

## 3. Presentation of the two perforated water aeration plates

*Version I:*

For the fine bubble generator previously studied [10] [11], there were 152 orifices with Ø = 0.1 mm; the air outlet area is:

$$A_{152} = n \cdot \frac{\pi \cdot d^2}{4} = 152 \cdot \frac{\pi \cdot \left(0.1 \cdot 10^{-3}\right)^2}{4} = 1.1932 \cdot 10^{-6} \text{m}^2 \tag{3}$$

Figure 6 shows the plate with 152 orifices.

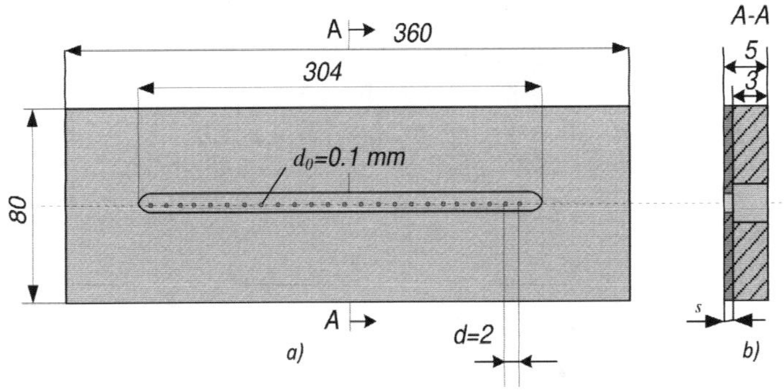

**Figure 6**. FBG orifices plate
a) plan view; b) cross section

Figure 7 shows the constructive solution of FBG for version I.
To make the orifices of Ø= 0.1 mm in the plate a channel with a depth of 3 mm and a length of 304 mm was created; the air outlet has a thickness of 2 mm. Subsequently, with the help of a special machine used for microprocessors type KERN Micro, the 152 orifices with Ø 0.1 mm were made in the channel.

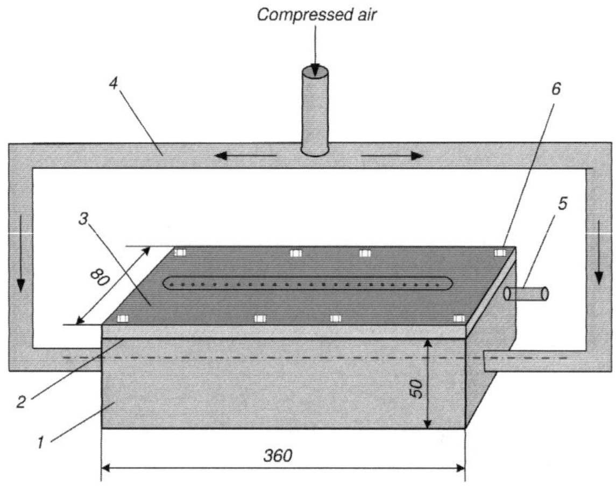


**Figure 7**. Fine air bubble generator.
1 - compressed air tank; 2 - sealing gasket; 3 – orifices plate; 4 - pipe with compressed air of the FBG; 5 - connection for measuring the air pressure in the tank; 6 - screws for fixing the perforated plate

*Version II:*
In the case of this second version, the orifices Ø =0.05 mm was made by micro-drilling with the help of two special machines used for micro processing, a drilling machine type Sixis (figure 8) on which the standard disc was made (figure 9) with 113 orifices Ø = 1 mm.

**Figure 8.** Sixis drilling machine

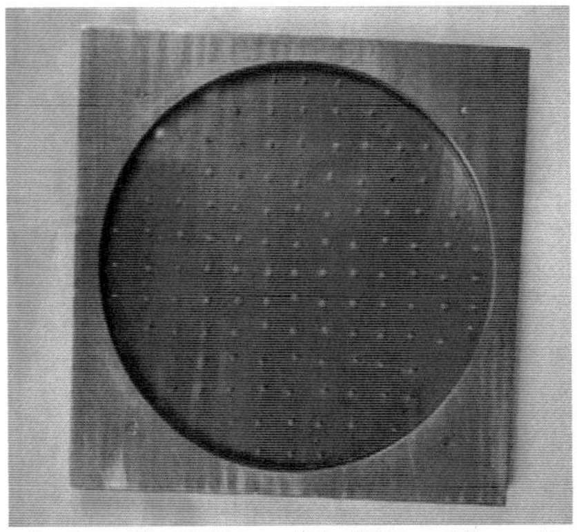

**Figure 9**. Standard disc with 113 orifices Ø =1 mm

The plates were fixed on standard discs and orifices Ø =1 mm was made leaving 0.3 mm of the thickness of the plates not perforated. The standard disc with the plates thus drilled were fixed on the Tripet MUS 100 machine (figure 10) where the orifices with Ø =0.05 mm (figure 12) were made on the last 0.3 mm unperforated, with the help of special Guhring drills (figure 11 a, b).

**Figure 10.** Tripet MUS 100 drilling machine.

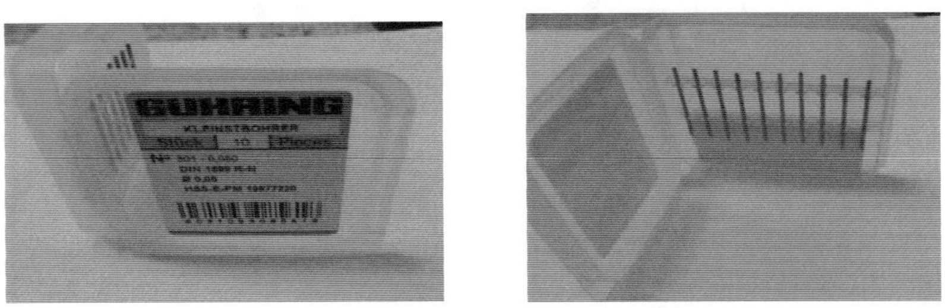


(a) (b)

**Figure 11**. Guhring drills Ø =0.05 mm.

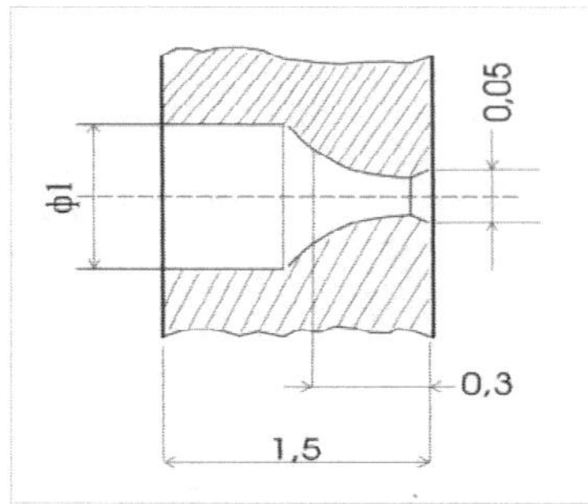


**Figure 12**. Making the orifices of Ø = 0.05 mm with the Tripet MUS 100 drilling machine.

For this version the plate has 113 orifices Ø = 0.05 mm (figure 13).
The air outlet area in the water will be given by the expression:

$$A_{113} = n \cdot \frac{\pi \cdot d^2}{4} = 113 \cdot \frac{\pi \cdot \left(0.05 \cdot 10^{-3}\right)^2}{4} = 0.221 \cdot 10^{-6}\ \text{m}^2 \tag{4}$$

Since in this version we have an installation with four perforated plates Ø = 0.05 mm, it results that the total air outlet area of the installation will be:

$$A_{113} = A_{i,t} = A_{i,4} = 0.221 \cdot 4 \cdot 10^{-6} = 0.887 \cdot 10^{-6}\,\text{m}^2 \tag{5}$$

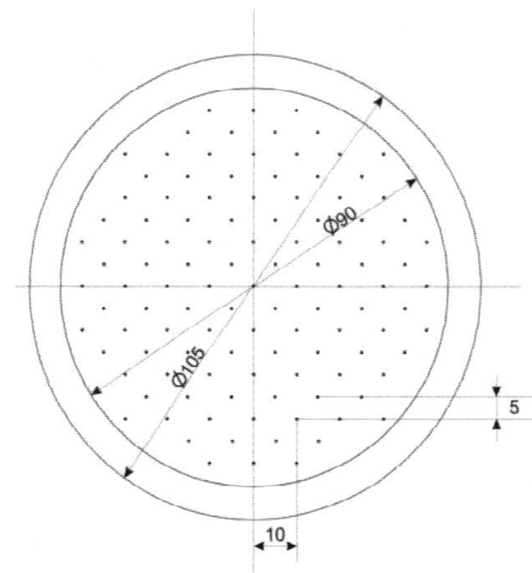


**Figure 13**. The orifice plate: n = 113; Ø = 0.05 mm.

The area of the air outlet section in the water will be:
In version I: $A_{152} = 1.1932 \cdot 10^{-6}\ \text{m}^2$
In version II: $A_{113} = 0.877 \cdot 10^{-6}\ \text{m}^2$
So: $A_{152} > A_{113}$ with: $\Delta A = (1.1932 - 0.877) \cdot 10^{-6} = 0.3062 \cdot 10^{-6}\ \text{m}^2$
In order to have the same section of compressed air outlet in the water (hence the same air flow rate), it is necessary to cover a number of orifices of the plate used in version I:

$$n = \frac{A}{\left(\pi \cdot d_0^2\right)/4} = \frac{0.3062 \cdot 10^{-6}}{\dfrac{\pi \cdot \left(0.1 \cdot 10^{-3}\right)^2}{4}} = 39\,orifices \tag{6}$$

In this case, the following will remain in operation: 152 - 39 = 113 orifices.
In conclusion:

$$A_I = n \cdot \frac{\pi \cdot d^2}{4} = 1134 \cdot \frac{\pi \cdot \left(0.1 \cdot 10^{-3}\right)^2}{4} = 0.887 \cdot 10^{-6}\mathrm{m}^2 \quad (7)$$

$$A_{II} = 0.887 \cdot 10^{-6}\mathrm{m}^2 \quad (8)$$

So, in the case of version I there is a plate with 113 orifices Ø = 0.1 mm, while in version II there will be 4 plates each with 113 orifices Ø = 0.05 mm.
Verification:

$$n_{FBG,I} = 113 \cdot \frac{\pi}{4} \cdot d_I^2$$

$$n_{FBG,II} = 113 \cdot 4 \cdot \frac{\pi}{4} \cdot d_{II}^2$$

$$\frac{n_{FBG,I}}{n_{FBG,II}} = \frac{1}{4} \cdot \frac{0.1^2}{0.05^2} = \frac{0.01}{0.01} = 1$$

## 4. The sketches of the two water aeration installations

a) For version I the sketch of the stand used for the experimental researches is presented in figure 14. The air flow rate and compressed air pressure were measured, the values remaining constant over time. In the case of this stand, the oxygenometer probe is a moving part with which the dissolved oxygen concentration in the water is determined, and the fine bubbles generator is fixed.

The measurement of the dissolved oxygen concentration in the water is done until the displayed value stabilizes.

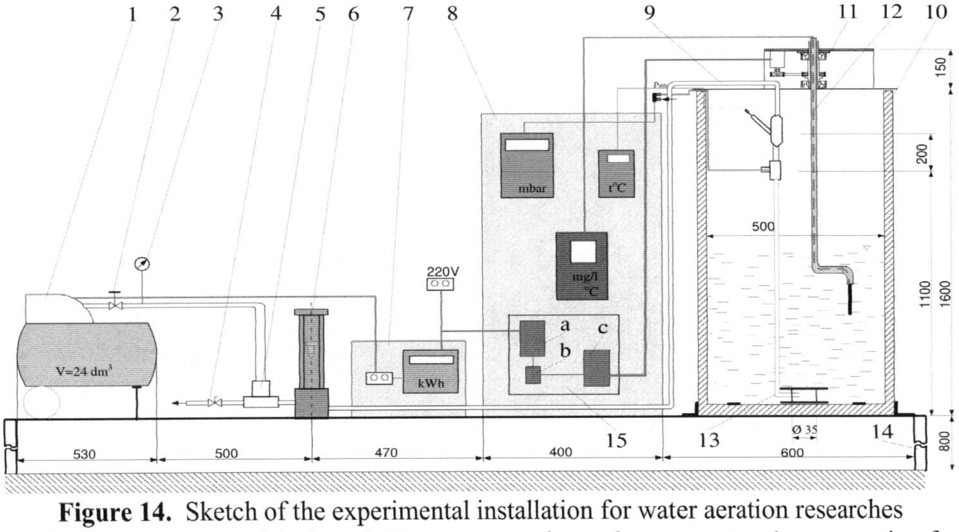


**Figure 14.** Sketch of the experimental installation for water aeration researches
1– electro compressor with air tank; 2– pressure reducer; 3-manometer; 4 – connection for evacuating air into the atmosphere; 5– T-joint; 6– rotameter; 7– electrical panel; 8– panel with measuring devices; 9– pipes for the transport of compressed air to FBG; 10– water tank; 11– probe actuation mechanism; 12– oxygenometer probe; 13– FBG; 14– support for installation; 15 – control electronics: a - power supply, b - switch, c - control element

For version II in figure 15, the sketch of the experimental installation designed to test a set of four microbubble generators is presented. The compressed air supplied by the electro compressor (2) is accumulated in the tank (3) and subsequently passes through the pressure reducer (4) to the assembly of four FBG's (7); the compressed air pressure must overcome the hydrostatic load (H) i.e., the height of the water layer in the tank (8) and the loss of air pressure when passing through the orifices [10][11][12].

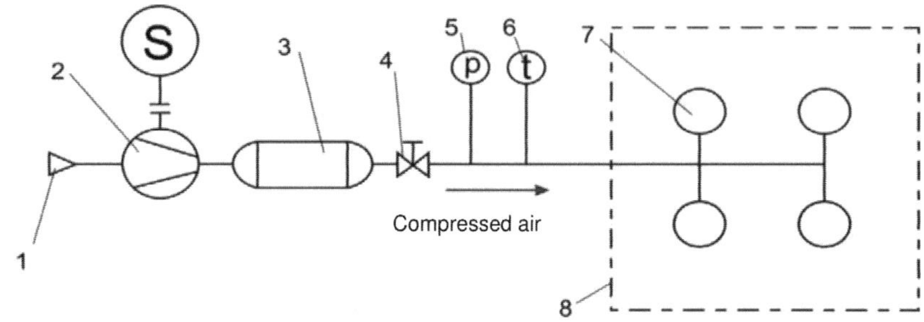


**Figure 15**. Sketch of the experimental installation for aeration of water with fine bubble generators. 1 - air filter; 2 - electro compressor; 3 - compressed air tank; 4 - pressure reducer; 5 - pressure measuring device; 6 - temperature measuring device; 7 - set of four fine bubble generators; 8 - the contour of the water tank

Figure 16 (a), (b) shows the set of four fine bubble generators.
Initial data for version I: $V_{H2O}$ = 0.125 [m$^3$], $\dot{V} = 600$ dm$^3$/h, $C_0$ = 5.84 [mg/m$^3$], $t_{H2O}$ = 23.7 °C; from [9] for this temperature results, a saturation concentration of dissolved oxygen in water of $C_s$ = 8.4 [mg / dm$^3$], $\tau$ = 120 min.

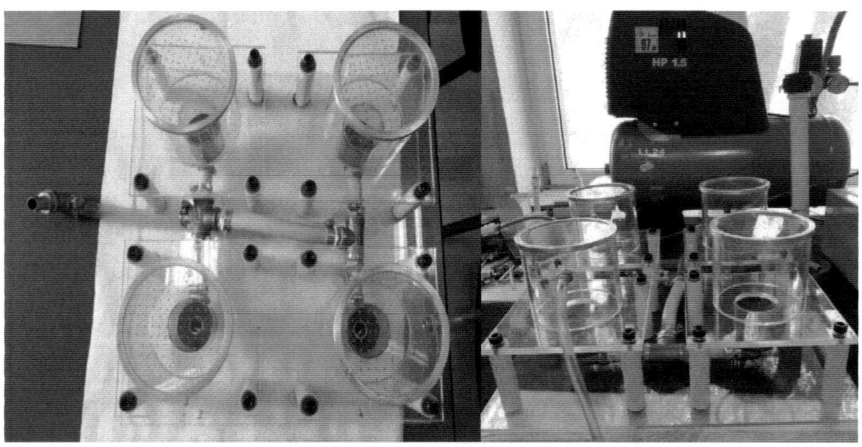

**(a)** **(b)**
**Figure 16**. Set of four fine bubble generators.

Table 2. Experimental values of concentration as a function of time for version I

| No. | $\tau$ [min] | 0 | 15 | 30 | 45 | 60 | 75 | 90 | 105 | 120 |
|---|---|---|---|---|---|---|---|---|---|---|
| 1 | $C_S$ [mg/dm$^3$] | 8.4 | 8.4 | 8.4 | 8.4 | 8.4 | 8.4 | 8.4 | 8.4 | 8.4 |
| 2 | $C$ [mg/dm$^3$] | 5.84 | 6.89 | 7.65 | 8.01 | 8.1 | 8.26 | 8.31 | 8.35 | 8.39 |

Based on the experimental data obtained, the graph of the function $C_{O2} = f(\tau)$ was drawn for version I: curve 1 in figure 17.
It is specified that on the ordinate from the graph in figure 17 there are two scales because on the graph paper the intervals:

$C_0 \rightarrow C_S$ version I: 5.84 $\rightarrow$ 8.4 = 2.56 mg / dm$^3$

$C_0 \rightarrow C_S$ version II: $5.84 \rightarrow 9.2 = 3.36$ mg / dm$^3$

They are contained in 50 mm.

For version II one can obtain:

50 mm ............... .3.36

3 intervals ............ x

---------------------------- x (an interval = 16.6 mm);

Each interval will be worth: $(3.36 \cdot 1.66) / 5 = 1.11$ mg / dm$^3$

It is obtained on the ordinate:

$y_1 = 5.84 + 1.11 = 6.96$;

$y_2 = 6.96 + 1.11 = 8.07$;

$y_3 = 8.07 + 1.11 = 9.18 \approx 9.2$ mg / dm$^3$.

As initial data for the experimental researches in case of version II there were specified as initial data: $V_{H2O}$ = [0.125 m$^3$], $C_0$ = 5.84 [mg / dm3], $p_{air}$ = 529.38 [mmH$_2$O], $t_{H2O}$ = 19.5 °C - from [9] for this temperature results, a saturation concentration of dissolved oxygen in water of Cs = 9.20 [mg / dm$^3$], $\tau = 0 - 120$ min.

Starting from the above data, the experimental researches was started as follows:

1) $C_0$ [mg / dm$^3$] was measured;
2) Air and water temperature were measured $t_{air}$ = 23.5 °C, $t_{H2O}$ = 19.5 °C.
3) The air flowrate supplied by the compressor was measured;
4) $C_{O_2}$ was measured every 15 ';
5) After two hours, the experiment was completed.

The results of the experimental measurements are presented in Table 3.

**Table 3**. Experimental values of concentration as a function of time for variant II.

| No. | τ [min] | 0 | 15 | 30 | 45 | 60 | 75 | 90 | 105 | 120 |
|---|---|---|---|---|---|---|---|---|---|---|
| 1 | *Cs* [mg/dm$^3$] | 9.2 | 9.2 | 9.2 | 9.2 | 9.2 | 9.2 | 9.2 | 9.2 | 9.2 |
| 2 | *C* [mg/dm$^3$] | 5.84 | 7.9 | 8.2 | 8.35 | 8.53 | 8.71 | 8.75 | 8.85 | 9 |

Based on the results recorded in table no. 3 the graph of the function $C_{O_2} = f(\tau)$ was drawn, for version II, curve 2 in figure 17.

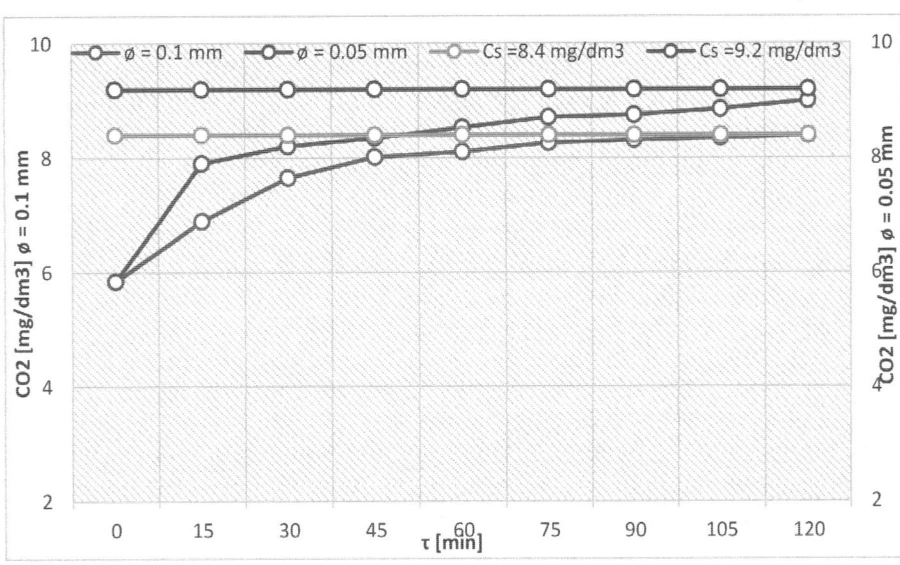

**Figure 17**. Graphical representation of the function $C_{O_2}$ = f(τ) for version I and for version II.

Comparing the values of C [mg / dm$^3$] in Table 2 with those in Table 3, one can observe that those in Table 3 are higher, so the aeration in version II is more efficient.

## 5. Conclusions

1) Figure 17 shows that the values of the dissolved oxygen concentration in water are higher in version II, curve 2 than those of version I, curve 1; on the axis of dissolved oxygen concentration in water there are two scales. The results of the experimental researches presented above are similar to those in the papers;
2) From figure 17 one can observe that the increase of the dissolved oxygen concentration in water from $C_0$ to $C_S$ is faster in case of using the fine bubble generator with bubbles having a diameter Ø = 0.05 mm. This fact confirms the theory that when reducing the size of the diameter of the bubbles immersed in water, the concentration of dissolved oxygen in water increases rapidly;
3) From the sketches of the experimental installation one can observe that this original solution is equipped with modern measuring instruments, of increased precision, with digital indication;
4) A special constructive solution consists in the construction of the driving mechanism of the oxygenometer probe which, during the measurements, rotates in the water tank with a constant speed of 0.3 m/s.

## References

[1] Akçaalan R, Köker L and Albay M 2016 Do eutrophic waters prompt to toxic cyanobacteria in Turkish Black Sea coast? *Journal of Environmental Protection and Ecology* **17(5)** 584-92
[2] Cao W and Zhang Y 2014 Removal of nitrogen (N) from hypereutrophic waters by ecological floating beds (EFBs) with various substrates, *Ecological Engineering* **62(1)** 148-52
[3] Chang N B, Wanielista M P, Xuan Z and Marimon Z A 2014 Floating treatment wetlands for nutrient removal in a subtropical stormwater wet detention pond with a fountain, *Developments in Environmental Modelling* **26(12)** 437-67
[4] Tudor I, Deak G and Tudor M 2015 Zooplankton distribution and dynamics in a Gorgova-Uzlina, Sontea-Furtuna, Rosu-Puiu and Matita-Merhei complex of lakes from the Danube Delta-Romania, *Journal of Environmental Protection and Ecology* **16(2)** 434-42
[5] Oprina G, Pincovschi I and Băran G 2009 *Hidro-Gas-Dynamics Aeration systems equipped with bubble generators*, POLITEHNICA PRESS Publishing House, Bucharest
[6] Rasha C 2017 *The influence of the architecture of fine bubble generators on the increase of the concentration of dissolved oxygen in water*, PhD Thesis, University POLITEHNICA of Bucharest
[7] Tănase E B 2017 *The influence of the instilled gas composition in water on the dissolved oxygen content*, PhD thesis, University POLITEHNICA of Bucharest
[8] Pătulea Al S 2012 *The influence of functional parameters and the architecture of fine bubble generators on the efficiency of aeration installations*, PhD thesis, University POLITEHNICA of Bucharest
[9] Mateescu G M 2011 *Hydro-gas-dynamics of fine bubble generators*, PhD thesis, University POLITEHNICA of Bucharest
[10] Tsuge H 2014 *Micro and Nanobubbles, Fundamentals and applications*, Pan Stanford publishing, University of California, Irwin
[11] Besnea D, Băran N and Mateescu G 2009 Using non-conventional technologies in order to build fine bubbles generators, *The Romanian Review Precision Mechanics, Optics & Mechatronics* **19** 31-6
[12] Călușaru I M, Băran N and Pătulea Al 2012 The influence of the constructive solution of fine bubble generators on the concentration of oxygen dissolved in water, *Advanced Materials*

*Research* **538-541** 2304-10

IOP Conf. Series: Earth and Environmental Science **960** (2022) 012020 doi:10.1088/1755-1315/960/1/012020

# Opportunities to improve road traffic on short-length road connections by correlating intersection traffic lights

**A Boroiu, E Neagu and A A Boroiu**

Department of Automotive and Transports, University of Pitesti, 1 Targul din Vale Street, 110040 Pitesti, Romania

E-mail: alexandru.boroiu@upit.ro ; ORCID ID: 0000-0002-6908-5546
elena.neagu@upit.ro ; ORCID ID: 0000-0001-6243-753X
andrei.boroiu@upit.ro ; ORCID ID: 0000-0001-7313-3398

**Abstract**. The paper aims to explore the possibilities of improving road traffic in the central area of cities characterized by a longitudinal arrangement of the street network, with application for the case of Pitesti, where the road network in the central area consists of two main roads arranged longitudinally, having one-way regulated traffic, interconnected by several streets. A special traffic problem is reported in the city center: on the main road connecting the two boulevards, the vehicle storage space is insufficient - because the distance between the two road intersections is too small and there is no correlation between the Green phases of traffic lights in the two intersections. The research, based on traffic measurements performed with DataFromSky software and micro-simulation traffic analyses performed with Vissim PTV software, indicated that the best solution is the partial or total correlation of the green time from the traffic light intersections that delimit the connecting road artery. As, almost exclusively, the works dedicated to the correlation of green light of traffic lights treat the problem only along the road arteries, this paper raises a special issue and reveals the possibility of simple solutions, by correlating the traffic lights at the intersections connecting the main arteries.

## 1. Introduction. Formulation of the problem

In recent decades, road traffic problems have become widespread in large cities in Romania, especially for areas that are important hubs for city traffic flows, and are becoming more and more acute [1], [2].

These road traffic problems are also manifested in Pitesti, which has a longitudinal configuration of the road network [3], given the city's layout on the terraces to the right of the Arges River, a road network that is separated in the central area by a very long pedestrian alley, which involves a certain peculiarity of road traffic.

The road network in the central area of the city consists of two main roads arranged longitudinally, having one-way regulated traffic, interconnected by several streets. However, there is also an important road artery, Eroilor Boulevard, and at the traffic-light intersections between these 3 important road arteries a special traffic problem is signaled: before the Green time for access from Republicii Boulevard to Eroilor Boulevard expires, vehicle access is no longer possible, because until the next road intersection the storage space of vehicles is insufficient - since the distance between the two road intersections is too small and there is no correlation between the Green phases of the traffic lights in the two intersections.

IOP Conf. Series: Earth and Environmental Science **960** (2022) 012020 doi:10.1088/1755-1315/960/1/012020

The present paper aims to identify possible solutions to solve the problem reported and, of these, the most advantageous solution.

## 2. Assessment of current road traffic in the central area of the city

For a first evaluation, the program indicated in [4] was used. It allows the evaluation of road traffic at any time from 06.00-22.00, for each of the 7 days of the week.

The indicator used is the speed of movement, on a 4-level scale:

- Green (fast traffic, so free traffic, at which the speed is limited not by the traffic conditions, but by the way of regulating the traffic in the respective area);
- Orange (low speed due to partially congested traffic);
- Light red (very low speed due to heavily congested traffic);
- Dark red (very low speed, even with stops, due to congested traffic);

The road network in the central area of Pitesti municipality is characterized by the fact that it has a longitudinal arrangement, being made up of two longitudinal road arteries on which the traffic is organized in one direction, and several connecting streets, which allow vehicles to change direction.

Thus, the main road connection between Republicii Blvd. and the Dumbravei-Armand Calinescu-Egalitatii road artery is Eroilor Blvd., where the arterial circulation capacity is insufficient for the existing traffic, because the storage length of this connection artery is too small. It is found that the traffic is free on the two longitudinal arteries with one-way traffic, instead on the Eroilor Blvd. connecting road artery the traffic is partially congested. The road traffic on the Eroilor Blvd. artery is regulated at both ends by traffic lights, and through the observations made on the spot it was found that they are not correlated with each other, neither in one direction nor in the other.

Considering that Eroilor Blvd. is the main road artery through which the traffic flows on the longitudinal axis of the city (the two road arteries with one-way regulated traffic) have access to the side neighborhoods (Razboieni, Banatului) - figure 1, as well as the visual observations regarding the characteristics of the road traffic entering this artery on Republicii Blvd., it was decided to carry out measurements in the two intersections that delimit the road artery and, based on them, to analyse road traffic, in order to find a solution to eliminate the signaled road congestion phenomenon.

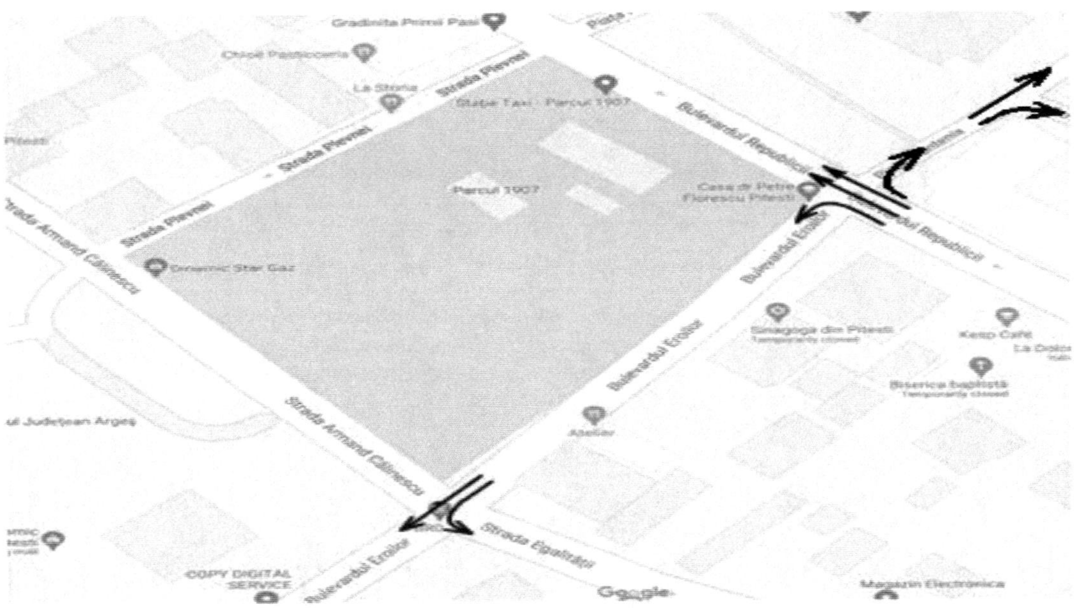


**Figure 1.** The direction of the road flows generated in Republicii Blvd. in the two traffic-light intersections that delimit the road section on Eroilor Blvd.

IOP Conf. Series: Earth and Environmental Science **960** (2022) 012020 doi:10.1088/1755-1315/960/1/012020

For this, traffic measurements were performed for the 2 intersections located at the confluence of the three targeted arteries: Republicii Boulevard, Eroilor Boulevard and Egalității Street, using the DJI Mavic Air Fly More Combo drone [5], and for the data analysis the associate "DataFromSky" software was used for processing traffic data. The two equipments represent an integrated system for collecting traffic data, with the help of which the analysis of road traffic can be performed for an area that includes several neighboring road intersections [6], [7].

Work was carried out raising the drone above the target area, at the required altitude, oriented in the optimal position, filming the traffic while the drone is maintained in the same position and processing of video captures made using the DataFromSky software, through which the following traffic data were obtained from each studied intersection:

- identification and metering of vehicles and pedestrians at the intersection (each moving object identified in the image is assigned a unique registration number);
- automatic distribution by vehicle categories (the program analyzes the filming and automatically classifies each moving object into 7 categories: bicycle, motorcycle, car, medium vehicle, heavy vehicle, bus, pedestrian);
- display of instantaneous values of speeds and accelerations / decelerations of vehicles;
- making graphs of vehicle speeds / accelerations in relation to time or space and displaying the average speed values;
- representation of the trajectory of each vehicle;
- making the Origin-Destination matrices for each vehicle identified (from which direction it entered the intersection and in which direction it moved after leaving the intersection);
- making colored traffic maps to highlight the density of vehicles on each arm of the intersection (from light blue for free traffic to dark red for congested traffic). Figure 2 presents the traffic density map made at the Republicii Boulevard - Eroilor Boulevard intersection in the time interval analyzed.

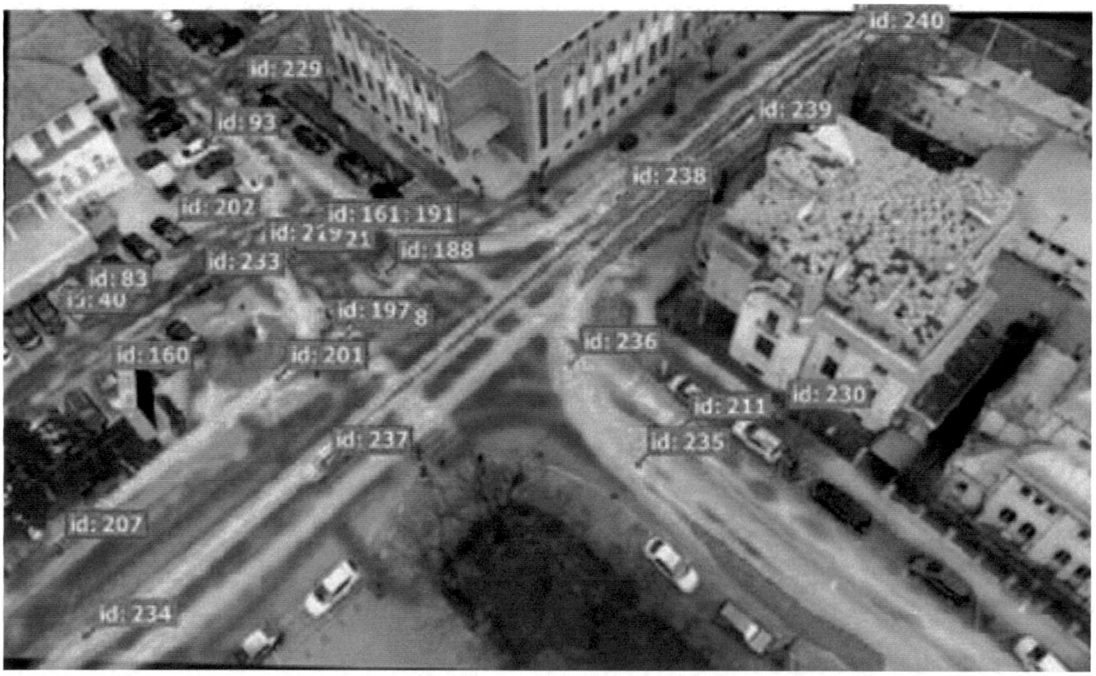


**Figure 2.** Traffic density map at the Republicii Boulevard - Eroilor Boulevard intersection.

By means of software one can export Excel reports on: traffic volumes, movement directions of vehicles entering the intersection, travel speeds, delay times, waiting lines length, these reports are very useful for simulating scenarios to improve traffic at the intersection.

After exporting the reports on the traffic volumes identified, the equivalence of physical vehicles into standard vehicles was performed, according to the SR 7348/2001 standard, Equivalence of vehicles for determining traffic capacity, which brought a series of additions and clarifications to the previous standard, STAS 7348/1986.

In this study, the coefficients for the equivalence of physical vehicles in $V_t$ standard vehicles were used, when the roads (including the streets) are level or have slopes of less than or equal to 2%, for streets in localities. After specifying these traffic conditions, based on the data collected by the special equipment (which counts the vehicles and identifies the category to which each metered vehicle belongs), the calculation of the equivalence of physical vehicles in standard vehicles $V_t$ is performed automatically with the well-known relation:

$$N = \sum_{i=1}^{n} N_i \cdot C_i \tag{1}$$

where:
N – traffic intensity expressed in standard vehicles hour;
$N_i$ – the number of physical category i vehicles, counted in the unit of time;
$C_i$ – the coefficient of equivalence for category i vehicles.

Based on the traffic measurements performed, the following reports were generated using the DataFromSky software:

- For the Republicii Blvd. - Eroilor Blvd. traffic-light intersection (Post Office intersection), where the entries and exits were noted according to figure 3, highlighting separately the flows towards the center and towards D-na Balasa Street.

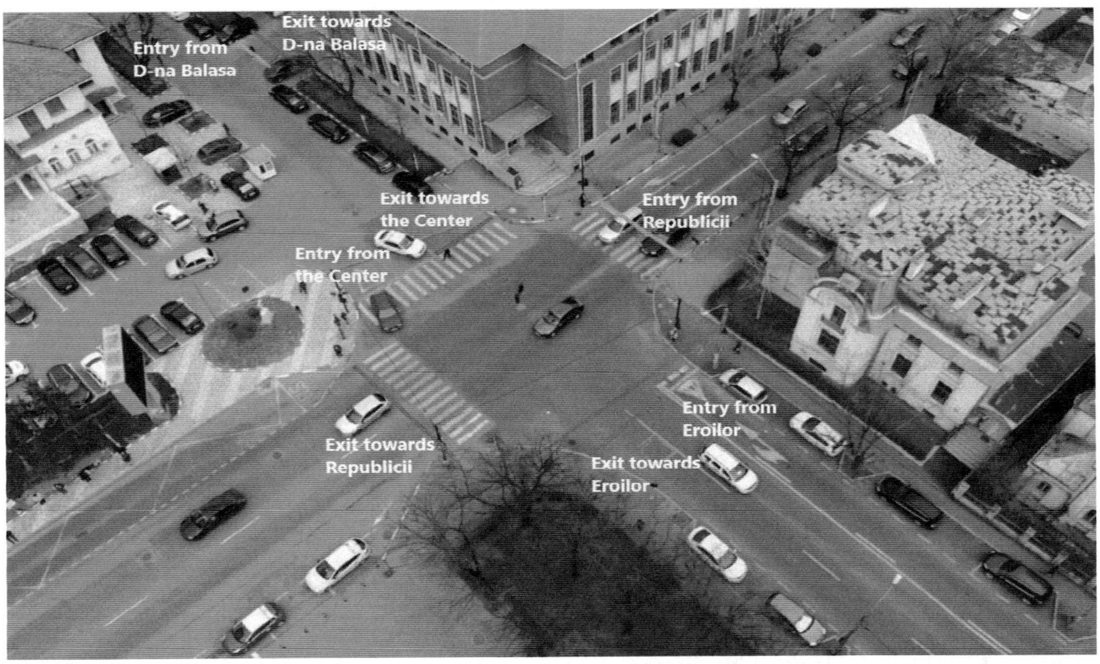


**Figure 3.** Notations for the movement directions of vehicles in the Republicii Boulevard – Eroilor Boulevard intersection.

IOP Conf. Series: Earth and Environmental Science **960** (2022) 012020 doi:10.1088/1755-1315/960/1/012020

The following values were obtained (the following data are specified in table 1and table 2: minimum identified speed, maximum identified speed, average speed, number of: cars, vans, trucks, buses, motorcycles, bicycles and total identified vehicles), after which the vehicles' equivalence, from physical vehicles to standard vehicles, was performed, while the cumulative value was rounded to the whole next number - both for the directions of entry into the intersection and for the directions of exit from the intersection.

**Table 1.** DataFromSky report for the intersection at the Post Office for the directions of entry.

| | Entry from Republicii | Entry from Eroilor | Entry from the Center | Entry from D-na Balasa |
|---|---|---|---|---|
| Min. speed (km/h) | 5.93 | 1.91 | 2.30 | 7.85 |
| Max. speed (km/h) | 61.04 | 25.66 | 27.67 | 29.25 |
| Average speed (km/h) | 26.23 | 14.32 | 13.74 | 16.80 |
| Cars | 1428 | 132 | 156 | 42 |
| Vans | 28,8 | 0 | 0 | 0 |
| Trucks | 0 | 0 | 0 | 0 |
| Buses | 126 | 42 | 0 | 0 |
| Motorcycles | 0 | 0 | 0 | 0 |
| Bycicles | 0 | 0 | 0 | 0 |
| Road traffic intensity (standard veh/h) | 1583 | 174 | 156 | 42 |

**Table 2.** DataFromSky report for the intersection at the Post Office for the directions of exit.

| | Exit towards Eroilor | Exit towards the Center | Exit towards Republicii | Exit towards D-na-Balasa |
|---|---|---|---|---|
| Min. speed (km/h) | 14.35 | 5.88 | 11.99 | 9.66 |
| Max. speed (km/h) | 26.51 | 35.79 | 59.65 | 13.22 |
| Average speed (km/h) | 20.23 | 14.61 | 34.96 | 11.76 |
| Cars | 516 | 276 | 972 | 36 |
| Vans | 14,4 | 14,4 | 0 | 0 |
| Trucks | 0 | 0 | 0 | 0 |
| Buses | 0 | 0 | 168 | 0 |
| Motorcycles | 0 | 0 | 0 | 0 |
| Bycicles | 0 | 0 | 0 | 0 |
| Road traffic intensity (standard veh/h) | 531 | 291 | 1140 | 36 |

- For the Eroilor Blvd. - Armand Calinescu Street traffic-light intersection (the intersection from the Museum), where the entries and exits were noted in the same mode, but according to the intersection, and for that the respective values were obtained.

Next, the level of service of each intersection was determined by calculating the average control delays using the DataFromSky software.

The level of service of a traffic-light intersection is determined by the aggregate control delays at the level of the entire intersection, presented in table 3 [8]:

For the Republicii Blvd. - Eroilor Blvd. traffic-light intersection, according to the reports generated with the help of the DataFromSky program, there resulted an average control delay of 33.7 s/veh, which classifies this intersection at a C-level of service.

For the Eroilor Blvd. - Armand Calinescu Street traffic-light intersection, according to the reports generated with the help of the DataFromSky program, there resulted an average control delay of 43.2 s/veh, which classifies this intersection at a D-level of service.

**Table 3.** Level of service for traffic light intersections.

| Level of service | Control delays (s/veh) |
|---|---|
| A | < 10 |
| B | 10- 20 |
| C | 20-35 |
| D | 35-55 |
| E | 55-80 |
| F | > 80 |

## 3. Identifying the causes of the problem

The traffic light intersection at the Post Office has a lower level of service than the traffic light intersection at the Museum, because the flow of vehicles entering from Republicii Blvd. to Eroilor Blvd. (the direction of the intersection) ends up creating long waiting lines, which even block the traffic light intersection at Republicii (figure 4) when their length is longer than the length of the road section between the two boulevards.

**Figure 4.** Waiting lines on Eroilor Blvd., from Republicii Blvd. to the traffic-light intersection at the Museum.

To identify the cause of the problem, the DataFromSky program can generate reports on the routes that each identified vehicle follows.

Based on the Origin - Destination report with the values recorded for each direction of movement in the case of the traffic light intersection at the Post Office, having the Museum intersection as destination, it can be noticed that the problem is generated by the high demand for access on Eroilor Blvd. of vehicles coming from Republicii Blvd. (408 vehicles / hour) and not by vehicles coming from the Center to Eroilor Blvd. (only 151 veh/h) - table 4 and table 5.

The vehicles coming from Republicii Boulevard and making a left turn to Eroilor Boulevard end up occupying the entire length of lane 1 on Eroilor Boulevard (preselected for the forward direction of movement) to the intersection, as this section is very short (about 115 m).

**Table 4.** Flows from the Post Office intersection - Republicii Entry.

| Eroilor Blvd. Exit | Exit towards the Center | Republicii Exit | Total |
|---|---|---|---|
| 408 | 204 | 876 | 1488 |

**Table 5.** Flows from the Post Office intersection - Entry from the Center.

| Eroilor Blvd. Exit | Exit towards the Center | Republicii Exit | Total |
|---|---|---|---|
| 151 | 0 | 29 | 180 |

Even if there are 2 lanes on Eroilor Boulevard, one of the lanes is destined to turn left, and the high traffic demand is forward, so that the left lane is almost empty and the forward lane is fully occupied, generating a length of the queue at peak hours that extends to Republicii Boulevard. It is obvious that the problem is generated by the vehicles coming from Republicii Boulevard.

## 4. Solutions to solve the problem

Based on what has been presented and considering that the hubs of the analyzed road network consist only of traffic-light intersections [9, 10], the following solutions are possible to solve the problem:

- modification of the traffic light cycle from the traffic light intersection at the Post Office, decreasing the total cycle time and, thus, the green time for vehicles coming from Republicii Boulevard (without reducing the absolute duration of the green time on an hourly interval), so that there would not be so many vehicles entering Eroilor Boulevard during a single green time phase, thus eliminating the risk of blocking the intersection at the Post Office;
- increasing the green time phase at the Museum traffic-light intersection for vehicles coming from Eroilor Boulevard, in order to reduce the length of the queue;
- the partial or total correlation of the green time from the Post Office traffic-light intersection (direction of movement Republicii Boulevard - Eroilor Boulevard) with the green time from the Museum traffic-light intersection (direction of movement Republicii Boulevard to Exercitiu Street.

However, the recommended solution is the one that will not affect in any way the capacities of the two traffic-light intersections, but will only solve the problem reported (the third solution), the correlation in green wave of the Green phase for forward and to the left direction of movement from the Republicii Boulevard with the Green phase for forward and to the left direction of movement from Eroilor Boulevard to Egalitatii Street.

This involves modeling the green time phase on the route containing the two intersections, where the main parameters are the type of movement (left turn made from Republicii Blvd. to Eroilor Blvd.), the distance between the two intersections (115 m) and the average travel speed – which, in this case, will be quite low.

The proposed solution presents sustainability for any of the 4 development scenarios for the Pitesti municipality, presented in the Sustainable Urban Mobility Plan of Pitesti [11], developed in 2017 and co-authored by the University of Pitesti.

## 5. Conclusions

The analysis of road traffic reveals that a specific problem of a road network with very close main arteries that are arranged in parallel, connected by very short road links, is the insufficient capacity of these connecting arteries for traffic flows generated in important nodes of the network.

For the case when these road nodes have the regulation regulated by traffic lights, the easiest solution can be the correlation of the Green phases from these intersections so that on the connecting artery not to create queues that would exceed the storage capacity of this artery, limited in mainly of its very small length.

There are many works that deal with the organization of road traffic in the "Green-Wave" system [12], [13], [14], [15], but the problem is treated almost exclusively along the arteries. Thus, the present paper raises a special problem and reveals the possibility of simple solutions, by correlating the traffic lights of the connecting intersections between the main arteries.

## References

[1] Boroiu A A and Neagu E 2015 *Trafic rutier si siguranta circulatiei rutiere. Aplicatii* (Pitesti: University of Pitesti Publishing House)
[2] Otat V and Otat O V 2017 *Analiza, modelarea si optimizarea circulatiei rutiere în zona Universitatii din Craiova* AGIR Bulletin no 4/2017 pp 16-21
[3] http://orasul.biz/harta-pitesti-ag/
[4] https://www.google.com/maps
[5] https://phantom.net/products
[6] Boroiu A A 2020 *Trafic rutier urban. Studii* (Pitesti: University of Pitesti Publishing House)
[7] Boroiu A A *Analiza traficului rutier. Aplicatii DataFromSky - PTV Vissim* 2021 (Pitesti: University of Pitesti Publishing House)
[8] CNADNR 2010 *Standard for the Development of Intersections at Public Roads* Road Technical Bulletin, no 4/2010 Designer Search Corporation
[9] Transportation Research Board (TRB) of National Academies of Science 2015 *Highway Capacity Manual* (Washington: National Academies of Science)
[10] Hall F L 1995 – *Fundamentals of transportation. Traffic stream characteristics* (Ontario: McMaster University
[11] Vîlcan A et al 2017 *The Plan of Sustainable Urban Mobility of the Municipality of Pitesti* Project of the Municipality of Pitesti www.primariapitesti.ro
[12] Romocea S, Ganea M and Albu A 2010 *An aproach of the simulation of "green light" in traffic using the gwtool software* Annals of the Oradea University, Fascicle of Management and Technological Engineering, no 3, vol 9, pp 139-143
[13] Wu X, Deng S, Du X and Ma J 2014 *Green-Wave Traffic Theory Optimization and Analysis* World Journal of Engineering and Technology, vol 2, no 3B, pp. 14-19, DOI: 10.4236/wjet.2014.23B003
[14] U.S. Department of Transportation Federal Highway Administration 2008 *Traffic Signal Timing Manual* Publication number FHWA-HOP-08-024
[15] Zhang X, Blockpoel R and Lu M 2019 *Adaptive green wave with speed advice for automated vehicles* 13-th ITS European Congress, Brainport, the Netherlands, 3-6 June 2019, Paper number ITS-1758

# Estimating the impact of a new bridge over the river Danube

**Adrian Vilcan[1], Viorel Nicolae[2], Ivana Martin[3], Mihaela ISTRATE[4]**

[1]AV TRANSPORT PLANNING SRL, Ionel Perlea 11, Apt. 15, S1, Bucharest, Romania, adrian@vilcan.com
[2]University of Pitești, Faculty of Mechanics and Technology, Târgu din Vale Street, no.1, Pitești, Romania, viorel.nicolae@upit.ro
[3]AV TRANSPORT PLANNING SRL, Ionel Perlea 11, Apt. 15, S1, Bucharest, Romania, ivanamartin.mi@gmail.com
[4]University of Pitești, Faculty of Mechanics and Technology, Târgu din Vale Street, no.1, Pitești, Romania, mihaela.istrate@upit.ro

E-mail: viorel.nicolae@upit.ro ORCID ID : 0000-0001-7598-7434

**Abstract**. The impact of a new bridge over the Danube was estimated using a 3-step road transport model, as at the time no multimodal model was available. Different forecasts have been considered for both scenarios without the project and with the project. Also, sensitivity tests have been carried out in order to assess the impact of various bridge tolls. At the end, time savings and emissions have been estimated.

## 1. Introduction

This paper is based on the application and implementation of the former Road Transport National Model in Romania in order to assess the impact and feasibility of a new bridge over the Danube in Braila [1].

Main model features are:

- Time dimension: AADT level (average day).
- Zoning system: 125 zones of which 54 zones for the study area (counties Braila, Galati, Tulcea).
- Road network: 1.486 links consisting of motorways, national roads, county roads, local roads, bridges and ferries.
- Five categories of vehicles: cars, buses, trucks 2 axles, trucks 3&4 axles, trucks 5+ axles.
- Generalised cost considered as impedance for assignment.
- Multiple route choice assignment calibrated by category of vehicle.
- Three stage model: Generation / Distribution / Assignment by category of vehicle.
- Software platform: VISUM.

The paper shows how a national road transport model can be detailed and recalibrated for the purpose of specific projects in order to obtain a reasonable estimation of the impact of implementing new infrastructure. Moreover, the paper shows how the model was calibrated for each step.

## 2. Model calibration

### *2.1. Model Trip generation/attraction*

Modelling trip generation/attraction has been tested with different combinations of population, GDP, GDP per capita, employment and car ownership. Finally the following approach has led to the best results.

$$Tg,m = xg,m * pop + yg,m * GDP/capita \quad (1)$$

$$Ta,m = xa,m * pop + ya,m * GDP/capita \quad (2)$$

Where:

- T - number of trips
- pop - total population
- x, y - regression factors
- a, g - attraction, generation
- m - mode

*2.2. Model Trip distribution*

The distribution model is considered to be the gravity one, described hereunder. The trip distribution model is represented by the equation:

$$Ti,j = ai.bj.Vi.Aj.F(Ci,j) \quad (3)$$

Where:

- Ti,j - trips estimated from zone i to zone j;
- ai, bj - factors to secure mathematical consistency;
- Vi - total number of trips generated by zone i;
- Aj - total number of trips attracted by zone j;
- F(Ci,j) - cost deterrence from zone i to zone j.

The deterrence function is represented by the equation:

$$F(Ci,j) = Ci,j.X1exp(X2.Ci,j) \quad (4)$$

Where

- X1, X2 - parameters of trip length distribution
- Ci,j - generalized costs for a trip from zone i to j

For the calibration of the distribution model the distance has been considered in a first instance as generalised cost. Thus, in this case Di,j has replaced the Ci,j. However, for the estimation of bridge induced traffic the distribution model has also been calibrated on the generalised cost – considering time and costs as influencing values.

The calibration process produces estimates of the coefficients a, b, X1 and X2 of the impedance function. The parameters are optimised by a maximum likelihood estimation procedure, using both parameters X1 and X2 or only one parameter X2.

*2.3. Model Assignment*

The assignment technique considered for each category of vehicles is the "stochastic assignment with logit multiple route choice". The impedance function for each category of vehicles "i" is given by:

$$IMPi = Timei \ x \ VOTi + Disti \ x \ VOCi + TOLLi \quad (5)$$

Where:

IMP = impedance

Dist = distance

VOT = Value of Time

VOC = Vehicle operating cost

TOLL = toll on specific links if is the case

The following VOT is considered for Romania, based on HEATCO and JASPERS studies, at the level of year 2010:

- o Passenger cars: 12.15 €/hour for one person
- o Buses: 9.75 €/hour for one person
- o Trucks 1.98 €/hour for one ton of freight

The following level of occupancy (persons per vehicle) is considered for passenger transport:

- o Passenger cars: 2.1
- o Bus 25.0

The following average loading factor is considered for the freight vehicles (tons per vehicle):

- o Truck 2 axles: 1.5
- o Truck 3&4 axles: 5.0
- o Truck 5+ axles: 11.0

Regarding the Volume/Delay functions, at this moment there are not available these functions for Romania at the level of national roads. Thus, the volume/delay function considered is the BPR1 function of VISUM transport planning software. This function is as follows:

$$tcur = t0 \cdot (1 + a \cdot satb) \quad (6)$$

Where:

sat = q/(qmax·c), and:

sat = level of saturation

q = flow of vehicles in PCU

qmax = maximum flow of vehicles (capacity)

a = 1, b = 2, c = 1.

## 3. Traffic forecasts

Figures below shows the traffic flows in vehicle per day, all vehicles, in a normal working day, and the OD relations that will us ethe bridge, in the scenario with bridge [3, 4].

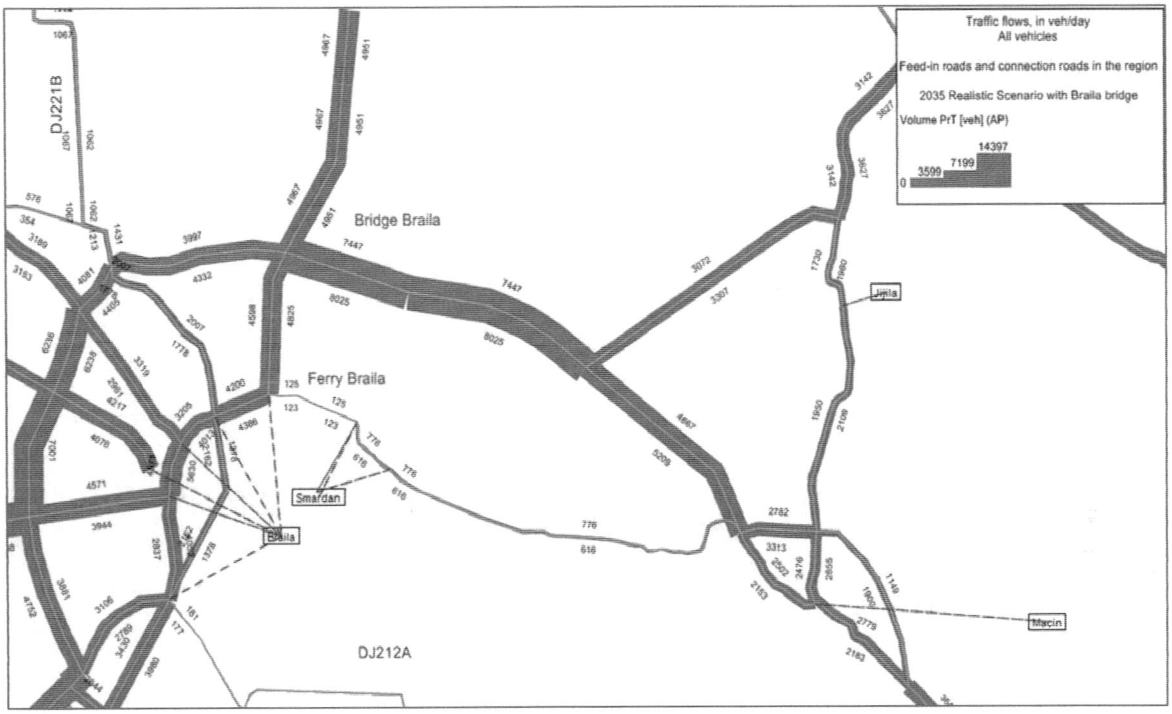


**Figure 1.** Traffic flows in veh/day, on the feed-in roads and connection roads – 2035 Realistic Scenario with bridge

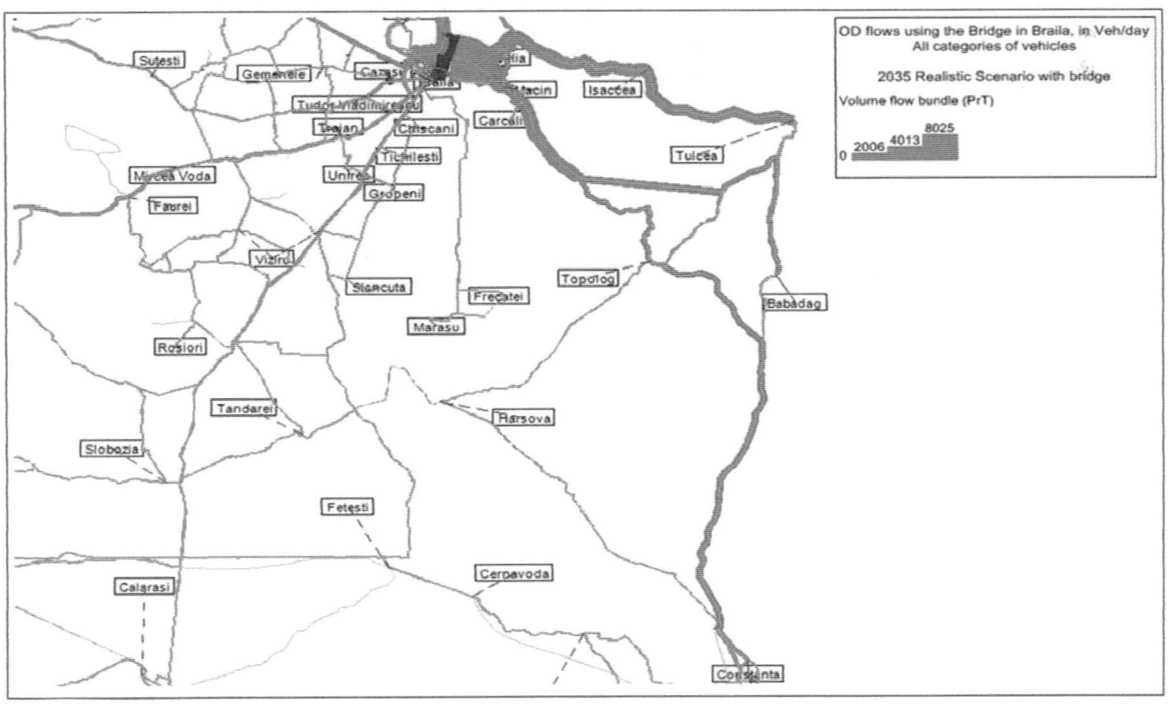


**Figure 2.** OD relations using the bridge in Braila – 2035 Realistic Scenario with bridge, core study area

Danube crossing trips are estimated as presented in the following tables [1].

**Table 1**. Estimation for Danube crossing trips in 2025.

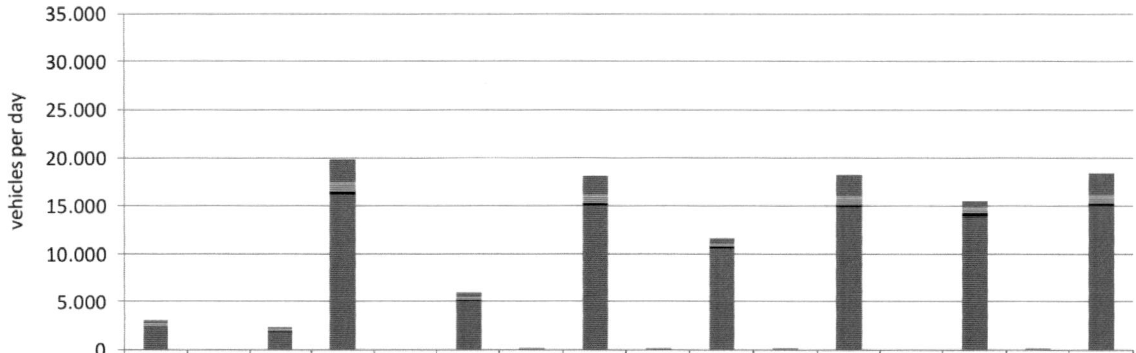


| | Business As Usual | | | | Pessimistic Scenario | | | | Realistic Scenario | | | | Optimistic Scenario | | | |
|---|---|---|---|---|---|---|---|---|---|---|---|---|---|---|---|---|
| | Braila Ferry | Braila Bridge | Galati Ferry | other | Braila Ferry | Braila Bridge | Galati Ferry | other | Braila Ferry | Braila Bridge | Galati Ferry | other | Braila Ferry | Braila Bridge | Galati Ferry | other |
| Total | 3.016 | | 2.368 | 19.834 | 43 | 6.033 | 250 | 18.209 | 183 | 11.646 | 273 | 18.272 | 47 | 15.577 | 224 | 18.404 |
| HT | 164 | | 142 | 2.412 | 0 | 460 | 7 | 1.992 | 4 | 597 | 8 | 2.217 | 0 | 668 | 7 | 2.254 |
| MT | 194 | | 101 | 237 | 11 | 124 | 60 | 211 | 123 | 62 | 80 | 228 | 27 | 275 | 72 | 232 |
| LT | 125 | | 101 | 645 | 23 | 234 | 30 | 724 | 42 | 226 | 33 | 627 | 18 | 406 | 23 | 629 |
| Bus | 31 | | 71 | 325 | 0 | 144 | 0 | 237 | 0 | 244 | 0 | 229 | 0 | 314 | 0 | 222 |
| Car | 2.502 | | 1.953 | 16.215 | 9 | 5.071 | 153 | 15.045 | 14 | 10.517 | 152 | 14.971 | 2 | 13.914 | 122 | 15.067 |

*Remarks: "Other" means the other 2 Danube crossings: bridges at Giurgeni-Vadu Oii and at Cernavoda.*
Notes: HT – Heavy Truck, MT – Medium Truck, LT – Light Truck

**Table 2**. Estimation for Danube crossing trips in 2045.

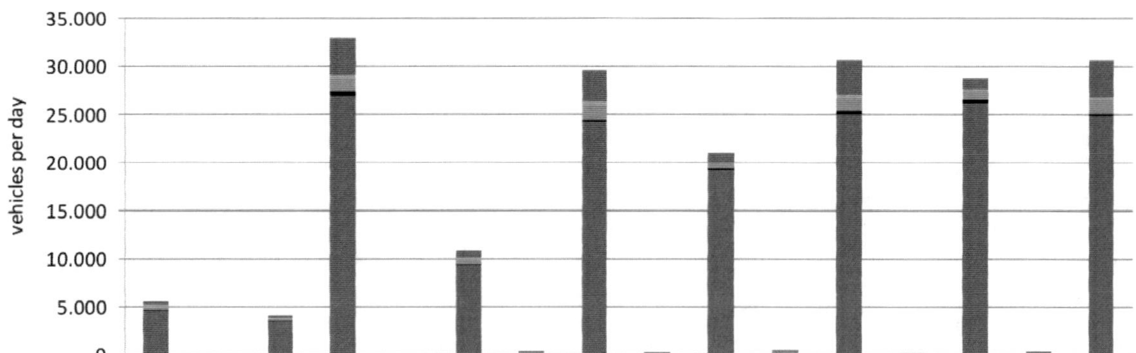


| | Business As Usual | | | | Pessimistic Scenario | | | | Realistic Scenario | | | | Optimistic Scenario | | | |
|---|---|---|---|---|---|---|---|---|---|---|---|---|---|---|---|---|
| | Braila Ferry | Braila Bridge | Galati Ferry | other | Braila Ferry | Braila Bridge | Galati Ferry | other | Braila Ferry | Braila Bridge | Galati Ferry | other | Braila Ferry | Braila Bridge | Galati Ferry | other |
| Total | 5.553 | | 4.132 | 32.958 | 73 | 10.977 | 439 | 29.680 | 340 | 20.998 | 606 | 30.706 | 279 | 28.837 | 438 | 30.671 |
| HT | 368 | | 191 | 3.843 | 0 | 793 | 11 | 3.326 | 10 | 924 | 49 | 3.611 | 0 | 1.094 | 12 | 3.725 |
| MT | 224 | | 118 | 339 | 16 | 186 | 89 | 301 | 200 | 91 | 123 | 342 | 201 | 295 | 120 | 346 |
| LT | 270 | | 202 | 1.342 | 44 | 486 | 68 | 1.510 | 93 | 483 | 72 | 1.335 | 76 | 850 | 50 | 1.336 |
| Bus | 40 | | 87 | 380 | 0 | 164 | 0 | 295 | 0 | 286 | 0 | 282 | 0 | 386 | 0 | 287 |
| Car | 4.651 | | 3.534 | 27.054 | 13 | 9.348 | 271 | 24.248 | 37 | 19.214 | 362 | 25.136 | 2 | 26.212 | 256 | 24.977 |

The travel time savings will increase, as shown below in the table below.

**Table 2.** Travel time savings, in a realistic scenario.

| Vehicle type | Travel time savings in 2025 [hours per day] | Travel time savings in 2035 [hours per day] | Travel time savings in 2045 [hours per day] |
|---|---|---|---|
| Cars | -8.202 | -11.260 | -17.754 |
| Buses | -165 | -195 | -237 |
| LT (Light Truck) | -187 | -271 | -499 |
| MT (Medium Truck) | -70 | -106 | -168 |
| HT (Heavy Truck) | -503 | -784 | -1.227 |
| **TOTAL** | **-9.127** | **-12.616** | **-19.885** |

The difference of driven mileage per day will increase, as shown below in the table below.

**Table 3**. The difference of driven mileage per day, in a realistic scenario.

| Vehicle type | The difference of driven mileage per day in 2025, [km per day] | The difference of driven mileage per day in 2035, [km per day] | The difference of driven mileage per day in 2045, [km per day] |
|---|---|---|---|
| Cars | -11.101 | -23.795 | -35.366 |
| Buses | -526 | -735 | -134 |
| LT (Light Truck) | 1.949 | 3.656 | 3.088 |
| MT (Medium Truck) | 330 | 531 | 500 |
| HT (Heavy Truck) | 1.547 | .587 | -16 |
| **TOTAL** | **-7.801** | **-18.756** | **-31.928** |

## 4. Conclusions

The infrastructure will be very competitive, due to the following:

1. Roads/ferry crossings in the neighbourhood of the bridge for securing the connection to Tulcea county and the northern part of Romania means that traffic flows on alternative crossings/routes are very limited, less than 3% of the traffic volumes on the bridge.
2. Roads/bridges currently used for long distance transit flows from Constanta to Braila and Galati as well as to the North, North-East, so the model analyses show limited traffic flows using Giurgeni-Vadu Oii and Cernavoda bridge for these trips.

So, construction of a two lane bridge_is recommended, considering that in 2045 around 21.000 veh/day (app. 24.500 PCU/day) are expected. The parameters foreseen are:

- peak hour of 15% implies 93 % volume/capacity ratio
- peak hour of 10% to 12 % implies 65% to 75% of volume/capacity ratio

Evaluation of feed-in roads shows that in 2035 the total traffic for all categories of vehicles is expected to be not higher than 5.200 veh/day for all feed-in roads
Finally, the reserve of capacity is sufficient on all feed-in roads, representing 7% reserve on ring road of Braila.

## References

[1] Vilcan A et al. 2012 *Update of feasibility study & preparation of traffic model for a new bridge over Danube* (TCS 30376) AV Transport Planning SRL / Komobile AT / ISEPP
[2] NEA Transport Research and Training bv for Senter Agency of the Dutch Ministry for Economic Affairs, 2004 *Development of the National Road Transport Model for Romania*
[3] * * * 2000 *Highway Capacity Manual* Institute of Transportation Engineers ISBN 0-309-06681-6
[4] * * * 1999 *Transportation Planning Handbook* Inst. of Transp. Engineers ISBN: 0-935403-33-7

# Electric bus platform for urban mobility

**E Tudor[1*], A Marinescu[2], R Prejbeanu[3], A Vintila[4], T Tudorache[5], D G Marinescu[6], D O Neagu[7], I Vasile[8] and I C Sburlan[9]**

[1] Lab Manager, senior researcher INCDIE ICPE-CA Bucharest
[2] ASTR VP, Craiova Branch
[3] Dr. Ing. at University of Craiova, Faculty of Automation, Computers and Electronics
[4] Department Head, INCD ICMET Craiova
[5] Professor, University POLITEHNICA of Bucharest
[6] Assistant professor University of Pitești
[7] General Manager INDAELTRAC SRL, Craiova
[8] Senior researcher, INCDIE ICPE-CA, Bucharest
[9] Senior researcher, INCDIE ICPE-CA, Bucharest

*E-mail: emil.tudor@icpe-ca.ro
ORCID : https://orcid.org/0000-0001-7186-2124

**Abstract**. Today, the technology of automatic battery charging based on Wireless Power Transfer (WPT) for the electric mass transit industry involving electric trains, buses and trams, is being used more and more. The modern solution described in this paper proposes an innovative technology for mixed charging of electric buses, either by wireless charging for 2-3 minutes in selected stations, or by plug-in charging at the end of the bus line, which results in only minimal energy storage on board - practically enough to get to the next charging station. The reduction of the weight of the battery packs determines the increase of the number of passengers transported, but also a reduction of the purchase price of the bus, without reducing the performances. The conversion can cost about half the price of new electric buses, depending on the condition of the vehicle and the extent of the work. This solution can be applied especially for the conversion of Diesel buses into electric buses which is not only sustainable, but also significantly better in terms of investment and operational costs, comparing with the purchase of new electric buses.

## 1. Introduction

In the EU, public transport consumes about 63% of fossil fuel and produces almost 30% of CO2 emissions [1,2]. Following analyses on air quality impairment through local NOx emissions and contribution to global warming [3] and Directive 2001/81/EC have imposed strict limits, control instruments and penalties on member countries.

Specifically, the fleet of buses with diesel engines is aged both physically and as a level of pollution as resulting from [4] which shows the situation of urban transport in Europe.

Introduction of battery-powered electric buses (EBs-Electric Bus Systems) or hybrid buses (as a transition solution) drastically reduce their carbon footprint. The incentives promote electric buses and have made China hosting nearly 400,000 e-buses and accounts for 98% of the global e-bus stock. North America and Europe are piloting and testing equipment, but also have significant heavy-duty clean

IOP Conf. Series: Earth and Environmental Science **960** (2022) 012022 doi:10.1088/1755-1315/960/1/012022

transportation targets for 2020 [5], 2025 and up to 2050. Europe is far behind, although it has the most restrictive legislation regarding air pollution.

## 2. The main features of High Power WPT Systems

### *2.1. Wireless Power installed on an EB and transmission distance Tx – Rx (Ground Guard)*

SAE J2954-2 [6] Task Force proposes a selection related to the power, as described in the following table. Although they refer to EBs with non-contact charging they are similar to those practiced at EBs plug-in and even diesel buses.

Installed power on an EB and transmission distance (ground clearance) Tx – Rx acc. [6].

**Table 1** Heavy Duty WPT Classes and Ground Clearance 250 to 350 mm acc. SAE J2954-2.

| WPT 5 | WPT 6 | WPT 7 | WPT 8 |
|---|---|---|---|
| 50 kW | 125 kW | 250 kW | 500 kW |

The distance between Tx and Rx at the inductive coupling is an important parameter of design of any WPT system. If at light EVs the ground clearance does not exceed 200 mm, in the case of EB it is usually between 250 and 350 mm. Sometimes an electromechanical system for lowering the Rx coil at a fixed distance, much lower than Tx (20 – 50 mm) is used to improve the inductive transmission at these relative distances.

The choice of the operating frequency of the DC - AC conversion is in the range of 19 – 25 kHz or 36 – 40 kHz, as decided by International Consensus and published in [7], [8] and [9].

#### *2.1.1. Li - Ion battery capacity used on Plug-In EBs.*

So far, for an urban Plug-In EB, the cost of the battery was about 40% of the total cost, and the weight of the battery represented over 20% of the total weight. Too much!

A typical example of plug-in EB, the “emil” bus presented in the lecture [10], consisting of an 8 m long bus for 50 passengers, with a battery of 220 kWh and a declared range of 300 km. For EBs, average energy consumption at constant speeds is about 2 - 2.5 kWh/km. The range of 300 km is not realistic (0.73 kWh/km), and it is even smaller than 100 km, to avoid deep discharge of batteries in urban mode. As a result, this EB cannot operate for 8 hours on an urban line without recharging

#### *2.1.2. Li – Ion battery capacity for a EB with opportunity charging along the route.*

WPT charging can remove the mentioned major obstacle above for plug-in urban EB. The proposed solution uses the so-called opportunity charging in certain stations on the route for a short period (2 - 3 minutes) and a charge of 10 - 15 minutes at the end of the line, as presented in [11]. At the end of the daily operation, there is a plugin charging system in the depot to restore the battery’s (SoC) to 100%.

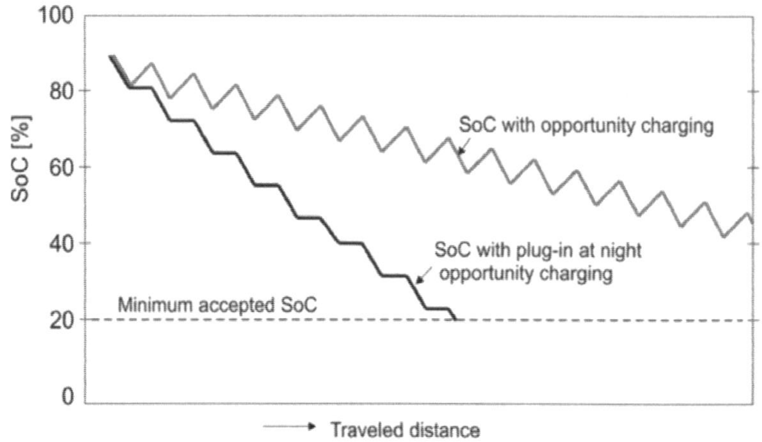


**Figure 1.** Traction battery's State of Charge with daily charging versus Daily and oportunity charging.

A short "opportunity charging" of 1-3 min, can extend the travelling duration by about 2-5 km. A daily route of 160 - 200 km can be covered by repeated charging on the route. For the same EB above the battery capacity can be only 25% that can be adapted in terms of weight and volume to fit the existing space in an older Diesel bus.

So, the opportunity charging to convert Diesel buses into EBs with a battery of much lower capacity as for plug - in EBs is an economical solution for Romania. As a result of wireless charging, the available energy described by the charging state (SOC) of a battery from such a WPT EB could be maintained using opportunity charging in an area of 40–60% SOC, which leads to an extension of battery life, instead of the large variation of the SOC within 20 to 90 % of the plug charging.

*2.2. Choosing the capacity of the battery pack*

The high voltage battery it is one of the most important component of a battery electric vehicle (EB). Choosing the battery pack is related mainly on the desired autonomy, on the imposed power of the motor and on the offers existing on the market.

Hence the R100 regulation regarding battery pack for EV, the batteries must be completed in packs called RESS (Rechargeable Energy Storage System) which must be homologated separately from the vehicle. The battery parameters have a significant influence on other components and attributes of the vehicle, like:

- maximum traction motor torque and maximum regeneration brake torque
- vehicle range, total weight and vehicle price

An example for a 12m bus with maximum energy consumption of 1,75 kWh/km and designed for a minimum range of 40 km, the battery must be more than 87.5 kWh.

$$E_b > e \cdot A/0.8 = 1750 \cdot 40/0.8 = 87.5\ kWh$$

For bigger buses a bigger consumption must be considered.

This consumption is related to full HVAC usage, when the autonomy can degrade by up to 40%. An agreed impact of the HVAC is considered to be 20%, achieved when the vehicle is set in "Economic Mode".

*2.3. Block diagram of electric bus with WPT charging*

Based on the experience gained in the development of the full wireless charging system for EV DACIA Electron, as being published in [11], it is proposed the block scheme in Fig. 2 which takes into account the following performance: inductively transferred power of at least 100 kW, ground clearance of 200 – 300 mm and proposed efficiency of power transfer at least 90%.

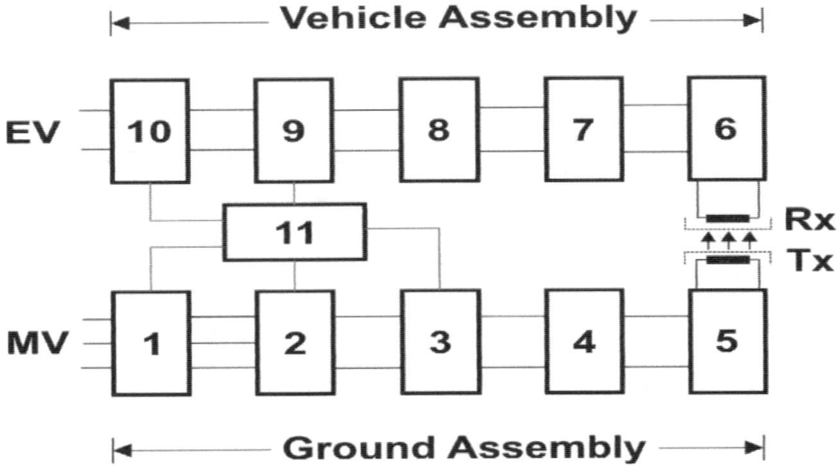


**Figure 2** Block diagram of a WPT drive for Electric Bus.

- 1- power-line filter, PFC, AC-DC Link with three-phase controlled rectifier and filter,
- 2- modular DC/AC inverter with operating frequency of 40 kHz,
- 3- impedance adapter,
- 4- primary LCC compensation circuit,
- 5- Tx ground block,
- 6- secondary LCC compensation circuit,
- 7- impedance adapter,
- 8- rectifier,
- 9- buck-boost converter,
- 10- EB battery,
- 11- microcontroller with Wi-Fi connections.

Imposed Performances

- Li-ion Battery: 88 kWh
- Power transferred min.100 kW
- Operating frequency: 40 kHz
- Ground clearance: max.300 mm

Energy transfer efficiency: >90%.

## 3. The main features of High Power WPT

The electric bus prototype with opportunity charging has a mechanical structure taken from a classic 8 m long bus equipped for power class WPT 6 (125 kW) proposed in the draft standard SAE J2945-2.

The system power supply is made from the three-phase electrical network in the 2x2 MIMO (Multiple Input – Multiple Output) modular system, i.e., it will have two Tx modules integrated into the road and two Rx modules located under the bus chassis. Each of Tx and Rx is sized for a transferred powers of 50 - 75 kW each and have identical dimensions (1x1m) being arranged next to each other in the travel direction (fig.3).

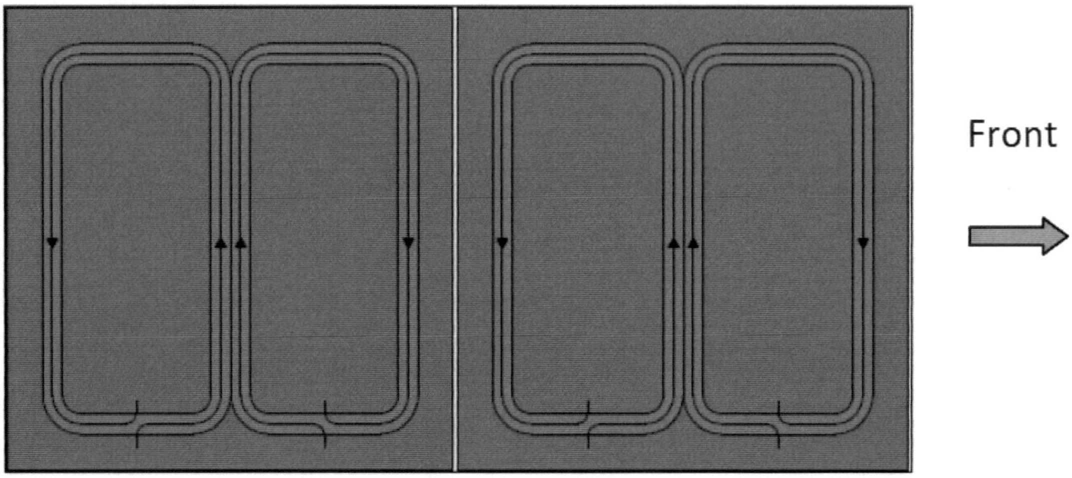


**Figure** 3 DD Tx modules (2MI).

Both couplers are bipolar double D (DD) coils, the solution that has the advantage as being able to provide the energy transfer for a ground clearance of at least 200 mm, double the value as opposed to circular planar coils. In this way it can stop using the classic electromechanical device for adjusting the distance during the batteries recharging process. For a ground clearance of 250 - 350 mm it can use the existing bilateral chassis lowering system. It should be noted that Tx is embedded in magnetic concrete according to the drive solution presented in the paper [12], which proposed the replacement of brittle ferrite magnetic field concentrators (figure 4).

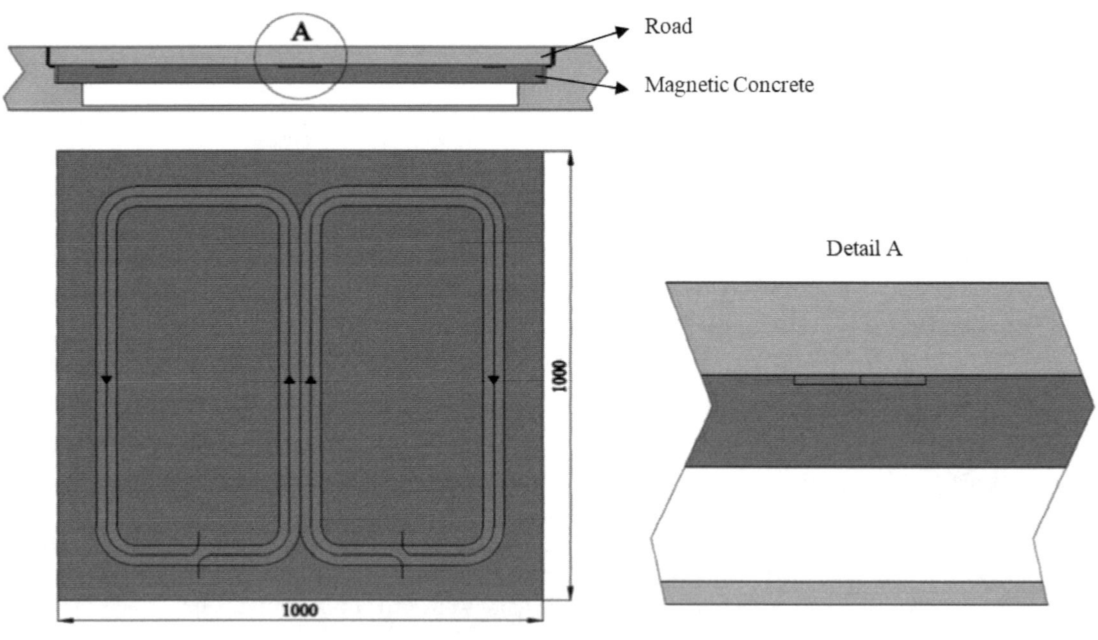


Figure 4 Tx module integrated in the road. (a)-section and top view, (b)-detail A.

The proposed 2x2 MIMO system adds an efficient LCL - LCL topology and a serial connection of the outputs of the two controlled rectifiers (fig.5). The arrows indicate signals from the sensors for the control loops.

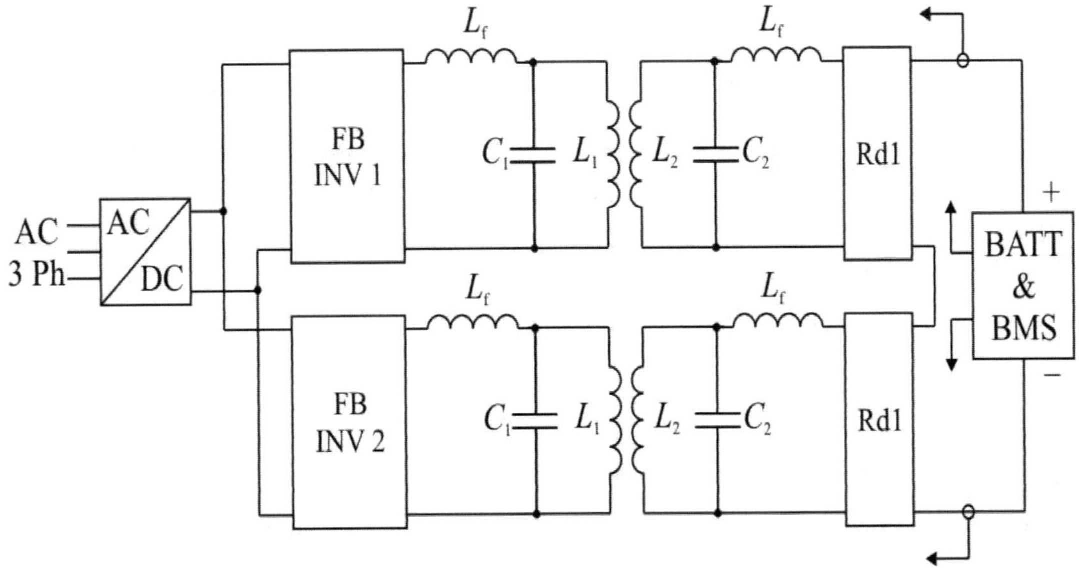


**Figure 5** Block diagram of the WPT 2x2MIMO system.

In this way we get a MIMO system where Tx and Rx modules operate independently under any conditions. The same system can be extended to higher powers, 4x4 MIMO or 6x6 MIMO i.e. up to 300 or 450 kW.

## 4. Conclusion

The successful conversion of an EV is based on some theoretical concepts presented here. Meanwhile, the constraints of the available materials are the most essential factors when choosing the elements of the newly powertrain. Most of the work are related to a proved and tested combination of transmission, motor, battery and power electronics. Still, for a new type of vehicle, as for improvements of the solution, new calculus must be applied by selecting new components and applying new features.

Electric bus has a better overall cost than the ICE, and is more reliable overall because there are fewer parts to fail. Most of the components are solid-state electronics with no moving parts. The engine of an electric car has a virtually infinite lifespan — the components will probably outlast the chassis. The only real expense is the batteries, which will need to be replaced about every 3 to 4 years.

The first part of the paper describes the current situation of the urban bus fleet in Europe and in the world and the transition imposed by the reduction of greenhouse gases to electric and hybrid buses (as a compromise solution).

An economical solution is proposed for the transformation of existing Diesel engine buses into electric buses based on the “opportunity charging” method achieved by WPT technology and the basic parameters of this conversion are established. In this way the capacity of the batteries and as a result the weight and the space occupied by them are substantially reduced.

## References

[1] Golovanov N and Marinescu A, Electromobility and Climate Change, *Proceeding of MPS (Modern Power Systems)*, 2019 Cluj Napoca, DOI: 10.1109/MPS.2019.8759786

[2] ****DIRECTIVE (EU) 2016/*2284 on the reduction of national emissions of certain atmospheric pollutants, *Official Journal of the European Union*, L 344/1,17.12.2016

[3] ***European Commission, *Air quality in cities,* http://ec.europa.eu/info/eu-regional-and-urbandevelopment/cities/priority-themes/air-quality-cities_en , Accessed 16 june 2020
[4] ****The European House on Eurostat data*, 2018 https://ec.europa.eu/eurostat/, Accessed 16 june 2020
[5] ***Wood Mackenzie editorial on 16.09.2019 The growing global e-bus landscape in China and beyond: *E-bus and infrastructure forecast for China, Europe and the US online*: https://www.woodmac.com/our-expertise/focus/Power--Renewables/ebus-landscape-2019
[6] ***SAE J2954-2, *Task Force Surface Vehicle Recommended Practice*, Wireless Power Transfer for Heavy-Duty vehicles,2019
[7] ****Wireless power transmission using technologies other than radio frequency beam*, ITU-R (ITU Radiocommunication Sector), REPORT SM.2303-2:2017
[8] ***IEC CISPR 11/Am 2/F1/Ed 6: 2017 *Candidate frequency bands also used for wireless power transmission (WPT) below 150 kHz*
[9] ****ISM (Industry, Scientific, medical) frequency bands*, https://en.wikipedia.org/wiki/ISM_band ,Accessed 14 july 2020
[10] Brandt F, “Primove”, *3rd Dresdner Fachkonferenz – Innovation Elektrobus*, December 2015
[11] Marinescu A, Vintila A, Marinescu D G, and Nicolae V., "Development of a wireless battery charger for Dacia Electron EV," *2017 10th International Symposium on Advanced Topics in Electrical Engineering (ATEE)*, 2017, pp. 241-247, doi: 10.1109/ATEE.2017.7905131.
[12] Marinescu A, Tudorache T, Vintila A and Dumbrava I, "A Comparative Assessment of Magnetic Concrete versus Ferrite for a High Power Inductive Coupler," 2021 *9th International Conference on Modern Power Systems (MPS)*, 2021, pp. 1-8, doi: 10.1109/MPS52805.2021.9492573

# Research on the influence of the clamping devices on the size of the perforated holes in long round pipes

**D I Poiană**[1] 0000—0002-4417-2803 **I Ionel**[2] 0000-0002-2999-1210 **and G Drăghici**[3] 0000-0002-3003-5015

Polytechnic University Timisoara, Faculty of Mechanical Engineering, Piata Victoriei no. 2, Timisoara, postal code 300006, Romania.

[1] dinupoiana@yahoo.com , dinu.poiana@student.upt.ro,
[2] ioana.ionel@upt.ro

**Abstract.**When perforation by cold plastic deformation processes of long round pipes, the technological problems appear primarily at the deformation of the part section, in the perforation area by losing the stability of the part, in the case of holes made at a distance from the pipe ends. The purpose of the research was to determine the influence of the constructive shape of the pipe clamping devices in the perforation area, in order to determine the device that offers the best stability of the section, to allow obtaining possible holes as large in diameter. The research was based on the application of materials strength methods and the Castigliano method of solving the indeterminacy, related to the process of cold plastic deformation. The research materialized by determining some calculation relations and specific diagrams regarding the diameter's variation of the perforated holes in the air environment, depending on the dimensions of the part and the mechanical characteristics of the part material, for different forms of clamping devices. The calculations were performed considering that the perforation is performed in air, with a punch applied from the outside of the considered piece. The calculation relations and the obtained diagrams allow the determination of the maximum orifices, for certain dimensions of the pipes and for different mechanical characteristics of the part material. At the conclusions, a comparison was made of the results obtained for the variants of clamping considered devices and the optimal shape of the device was established, which allows the making of the largest perforated holes in the part.

## 1. Introduction

Cold plastic deformation perforation is widely used to obtain holes in tubular parts. Long tubular parts are used as components in machine construction and in technical fields such as chemical industry, automotive industry, aeronautical industry, food industry, etc. Long tubular parts used in car construction are exemplified in figure 1, below.

Many of the long tubular parts must be provided with certain holes arranged along the tubular parts, at a distance from the ends, in which case technological problems arise regarding the execution of the holes by perforation.

For the perforation of the tubular parts, several perforation processes can be used in air [1- 4].

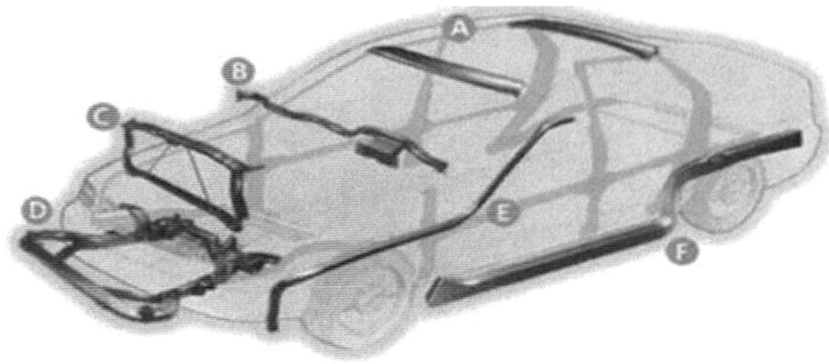


A. Roof Headers
B. Instrument Panel Supports
C. Radiator Supports
D. Engine Cradles
E. Roof Rails
F. Frame Rails

**Figure 1.** Tubular parts positioned at the car body, according to [1].

In the present paper is considered the case of perforating the pipe at a distance from the ends of the pipe, with single-edged tools, punch type, applied from the outside of the pipe, in air.

For this case, a theoretical research was carried out in this paper to determine the optimal shape of the pipe clamp device in the area of the perforated hole, in terms of diameters as large as possible to be made, taking into account the dimensions of the pipe (diameter and thickness of the pipe wall) and the mechanical characteristics of the pipe material (mechanical breaking strength and allowable mechanical strength).

The research is briefly presented in the following sections.

## 2. Research considerations

Three possible shapes for the round pipe clamping device in the area of the perforation hole were considered, according to figure 2, where cross sections are presented in the devices, through the center of the perforated hole [5-7]:

- a device marked with "Device 1", with the placement of the pipe on a flat plate and also pressed with a flat plate (figure 2 a);
- a device marked with "Device 2", with the placement of the pipe on a on a semi-cylindrical plate and pressed with a flat plate (figure 2 b);
- a device marked with "Device 3", with the placement of the pipe on a on a semi-cylindrical plate and pressed with a semi-elliptical plate (figure 2 c);

In order to perform the necessary calculations, the following were considered:

- the holes are made by the perforation process by cold plastic deformation, in air environment, with singular cut, of punch type;
- the necessary calculations are performed to determine the maximum diameter of the holes to be drilled, so that the piece does not deform in the drilling area; is determined for each type of device, Device 1, Device 2 and Device 3;
- the upper pressing plate will also be the punch guide plate;
- the length of the clamping device is considered sufficient of 80 mm, centered towards the hole to be made;

## 3. Determining the interdependence relationships between the diameters of the perforated holes and the dimensions of the pipes and the strength characteristics of the pipe material

### *3.1. Determining the stress state and raising the indeterminacy, for each of the 3 types of clamping devices.*

Because the holes in the work-piece are made by a plastic deformation process, at the moment of impact of the punch with the work-piece, tensions appear in the work-piece, due to the shear force that appears and causes the rupture and detachment of the material subjected to the perforation process [8,9].

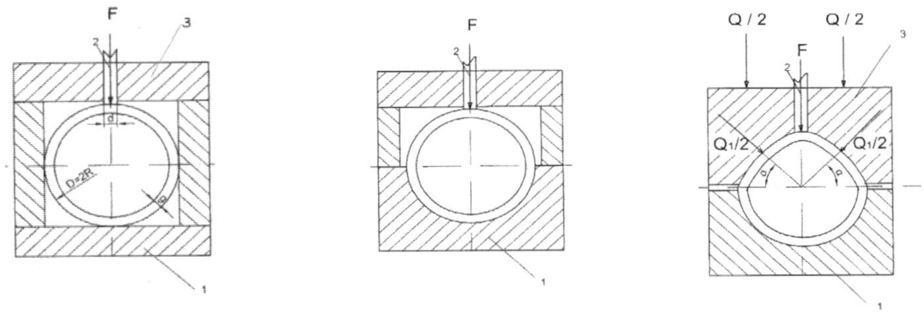


a b c

**Figure 2.** Device for clamping the tubular part in the hole area. Notations: 1 - the lower plate of the clamping piece; 2 - the perforation punch; 3 – the upper plate of the part clamping device and the punch guide; F - the actuating force of the punch; D - Outer diameter of the part; g - the thickness of the part wall; d - hole diameter / punch diameter

The stresses that appear in the material are determined, at the moment of the action of the punch on the part, based on the problems in the resistance of the materials, with analytical calculation relations and constructing the equilibrium schemes of forces and moments for the considered cases.
Figure 3 shows the force equilibrium scheme for each type of device,

- figure 3 a, for device 1;
- figure 3 b, for device 2 and device 3;

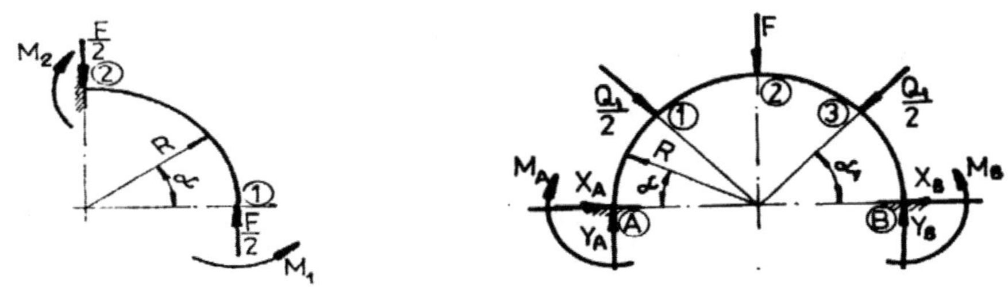


a - suitable for device 1 b - suitable for device 2 and 3

**Figure 3.** Equilibrium scheme of forces and moments

*3.2. Determining the calculation relationships of the maximum holes diameters*

*3.2.1. For the case of the device 1.* Castigliano's theorem is applied to remove the indeterminacy for the scheme in Figure 3. In section 1 we consider the zero rotation and obtain the relation:

$$\varphi_1 = \frac{\partial u}{\partial M_1} = \frac{1}{EI}\int M_i \frac{\partial Mi}{\partial Mi} ds = 0 \tag{1}$$

The bending moment in zone 1-2 will have the expression

$$M_{1-2} = M_1 + \frac{F}{2}R(1-\cos\alpha) \tag{2}$$

where F is the punching force

Replacing relation (2) in relation (1) results after performing the calculations:

$$M_1 = -\,0{,}18\ FR \tag{3}$$

$$N_{1-2} = -\frac{F}{2}\cos\alpha$$

$$T_{1-2} = -\frac{F}{2}\sin\alpha$$
$$M_{1-2} = (0{,}32 - 0{,}5\cos\alpha)FR \tag{4}$$

The stress diagrams for variant 1 are shown in figure 4

$$N_2 = 0$$
$$T_2 = -\frac{F}{2} \tag{5}$$
$$M_2 = 0{,}32\ FR$$

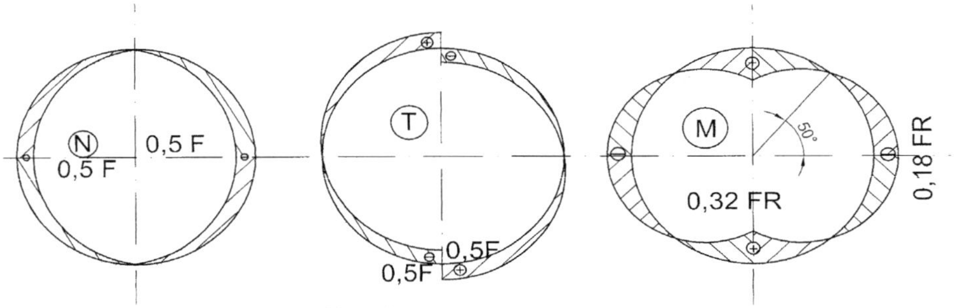


**Figure 4.** Effort diagrams for the case of the device 1.

By performing the calculations, the relation is obtained:

$$d \le \frac{\ell g}{1{,}91\pi R} \cdot \frac{\sigma_{ai}}{\sigma_r} \tag{6}$$

where: $\ell$ = the length of the clamping device; R = outer radius of the pipe section; g = pipe wall thickness; $\sigma_{ai}$ = permissible bending strength of the pipe material; $\sigma_r$ =limit breaking strength of the pipe material

*3.2.2. For the case of the device 3.* Castigliano's theorem is applied, for the scheme presented in figure 2.c, corresponding to the case when the piece is placed on a plate 1 with the semi-cylindrical surface and tightened with the semi-elliptical upper plate. The displacement $u_A$ and the rotation $\varphi_A$ are considered to be zero in section A. The system is obtained:

$$u_A = \frac{\partial u}{\partial X_A} = \sum_i \int \frac{M_i}{EI} \cdot \frac{\partial M_i}{\partial X_A} ds = 0 \tag{7}$$
$$\varphi_A = \frac{\partial u}{\partial M_A} = \sum_i \int \frac{M_i}{EI} \cdot \frac{\partial M_i}{\partial M_A} ds = 0$$

it can write the system:

$$X_A = X_B$$
$$Y_A = Y_B = \frac{F}{2} + \frac{Q_1}{2}\sin\alpha_1 \tag{8}$$
$$M_A = M_B$$

Considering that the force $Q_1$ acts at an angle $\alpha_1 = 45^o$ :

$$X_A = X_B = 0{,}46\ F + 0{,}02\ Q_1$$
$$Y_A = Y_B = 0{,}5\ F + 0{,}35\ Q_1 \tag{9}$$

$$M_A = M_B = 0{,}11\ FR - 0{,}025\ Q_1R$$

Figure 5 shows the effort diagrams corresponding to this case. In dangerous section 2 of the piece, the efforts have the values:

$$N_2 = -\,0{,}46\ F \qquad T_2 = 0{,}5\ F \qquad M_2 = 0{,}15\ FR. \tag{10}$$

The normal stress $\sigma_2$ in the dangerous section 2 of the part has the following expression:

$$\sigma_2 = \frac{F}{\ell g}\left(0{,}46 + 0{,}91\frac{R}{g}\right) \le \sigma_{ai} \tag{11}$$

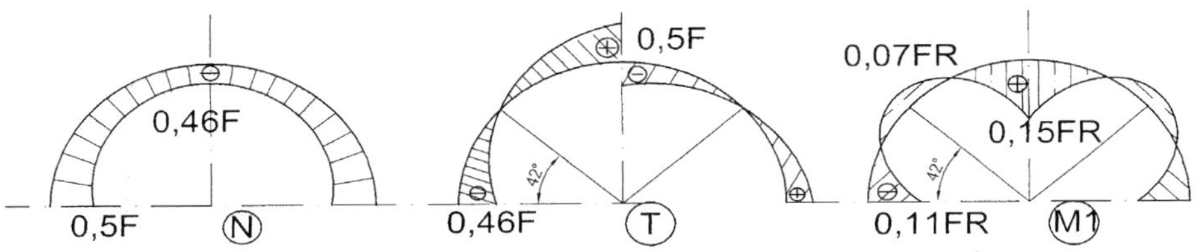


**Figure 5.** The effort diagrams corresponding to device 3.

Performing replacements and calculations, the calculation relationship of the diameter "d", is obtained which can be achieved by punching with a die without a cutting plate, when the pipe is placed on a semi-cylindrical plate 1 and is tightened with a semi-elliptical plate at the top, so:

$$d \le \frac{\ell g}{\pi(0{,}46g + 0{,}91R)} \cdot \frac{\sigma_{ai}}{\sigma_r} \tag{12}$$

where: $\lambda$ = the length of the profiled plate on which the pipe rests; R = outer radius of the pipe section; g = pipe wall thickness; $\sigma_{ai}$ = permissible bending strength of the pipe material; $\sigma_r$ = limit breaking strength of the pipe material

*3.2.3. For the case of the device 2.* For the device 2, it is a particular case of the situation with device 3, when there is no pre-tightening of the pipe on the upper plate of the device. The value of the force Q1 with which the plate 3 (figure 2 b) acts on the pipe, is determined from the condition of resistance of the pipe to the stress F (F = 0).

$$N_1 = -\,0{,}26\ Q_1 \qquad T_{1st} = 0{,}24\ Q_1 \qquad T_{1dr} = 0{,}26\ Q_1 \qquad M_1 = 0{,}06\ Q_1\ R \tag{13}$$

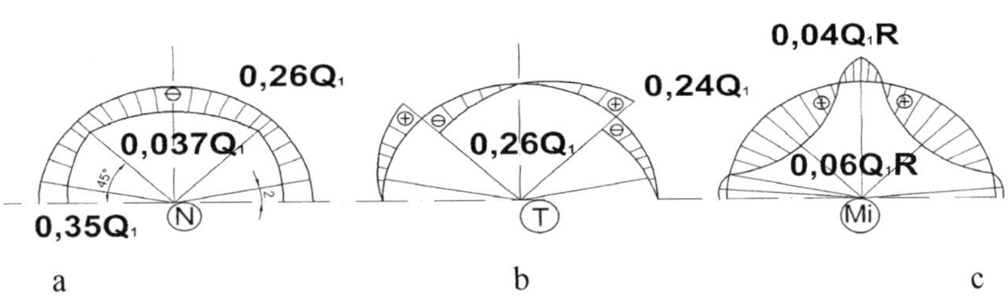


**Figure. 6.** Effort diagrams for the case of the device 2.

The effort diagrams are presented in figure 6.

$$\sigma_1 = \frac{Q_1}{\ell g}\left(0{,}26 + 0{,}4\frac{R}{g}\right) \leq \sigma_{ai} \tag{14}$$

$$Q_1 = \frac{\ell g^2 \sigma_{ai}}{0{,}26g + 0{,}4R} \tag{15}$$

$\alpha_1 = 45^o$:

$$Q_1 = 1{,}41\ Q \tag{16}$$

$$\alpha_2 = \frac{0{,}46F + 0{,}37Q_1}{\ell g} + \frac{6R}{\ell g^2}\left(0{,}15F - 0{,}08Q_1\right) \leq \alpha_{ai} \tag{17}$$

By performing the calculations, the relation (17) is obtained the maximum diameter "d" of the hole that can be made by punching when the pipe is placed on a semi-cylindrical plate and is tightened with a flat plate at the top:

$$d \leq \frac{\ell g}{\pi}\left(\frac{1}{0{,}46g + 0{,}91R} + \frac{0{,}5 - 0{,}75g}{0{,}25g^2 + 0{,}85gR + 0{,}72R^2}\right)\frac{\sigma_{ai}}{\sigma_r} \tag{18}$$

where: $\lambda$ = the length of the profiled plate on which the pipe rests; R = outer radius of the pipe section; g = pipe wall thickness; $\sigma_{ai}$ = permissible bending strength; $\sigma_r$ = limit breaking strength of the pipe material

## 4. Determining the variation diagrams of the size of the perforated holes, depending on the dimensional characteristics of the pipe and the mechanical resistance of the perforated pipe material

### *4.1. For the case of the device 1*

Based on the relation (6), a table of values was drawn up and the curves in figure 7 were drawn, in case the piece to be perforated is placed on a device with a flat surface and tightened with a flat plate.

The curves are suitable for long pipe type parts, with dimensional characteristics:

- outer diameters of the parts: D = 15, 20, 25, 30, 40, 60 mm
- the wall thickness of the pipes for each diameter considered: g = 1 ... 7 mm
- the length "$\lambda$" = 80 mm, material strengths: $\sigma_r$ = 330 MPa and $\sigma_{ai}$ = 150 MPa

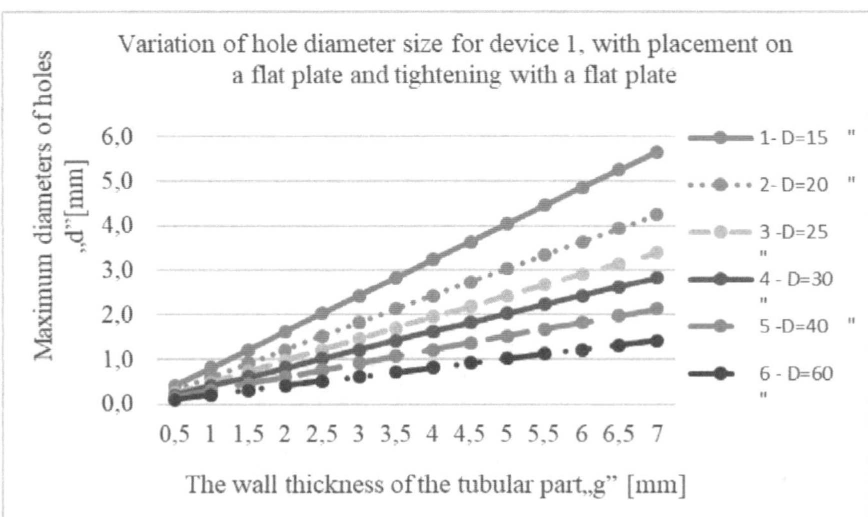


**Figure 7.** Diagram of variation of the diameters of the perforated holes, by using the device 1, with placement on a flat plate and tightening with a flat upper plate.

*4.2. For the case of the device 2*

Based on the relation (18), a table of values was drawn up and the curves of variation of the diameters were drawn, from figure 8, for the case when the perforated piece is placed on a device with a semi-cylindrical plate and tightened with a flat upper plate.

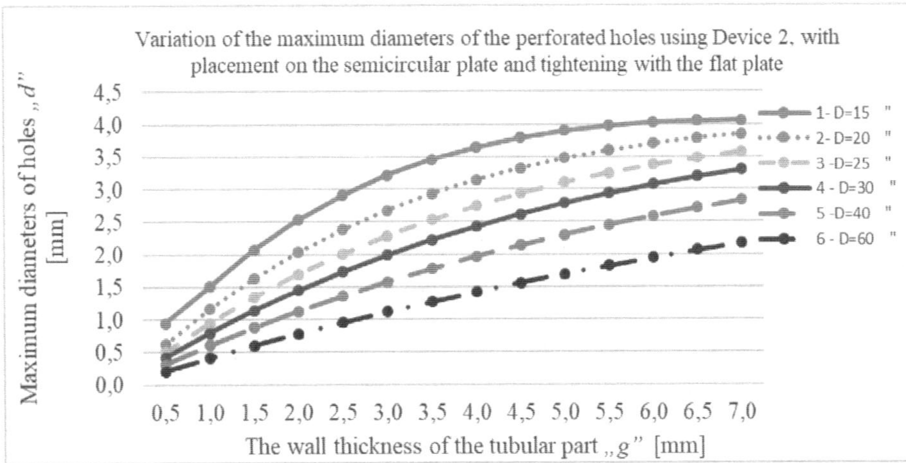


**Figure 8.** Diagram of diameters variation, by using the device 2, with placement on a semi-cylindrical plate and tightening with a flat upper plate.

*4.3. For the case of the device 3*

Based on the relation (12), a table of values was drawn up and the diameter variation curves were drawn, from figure 9, for the case when the perforated piece is placed on a device with a semi-cylindrical surface and tightened with an upper plate with semi-elliptical surface

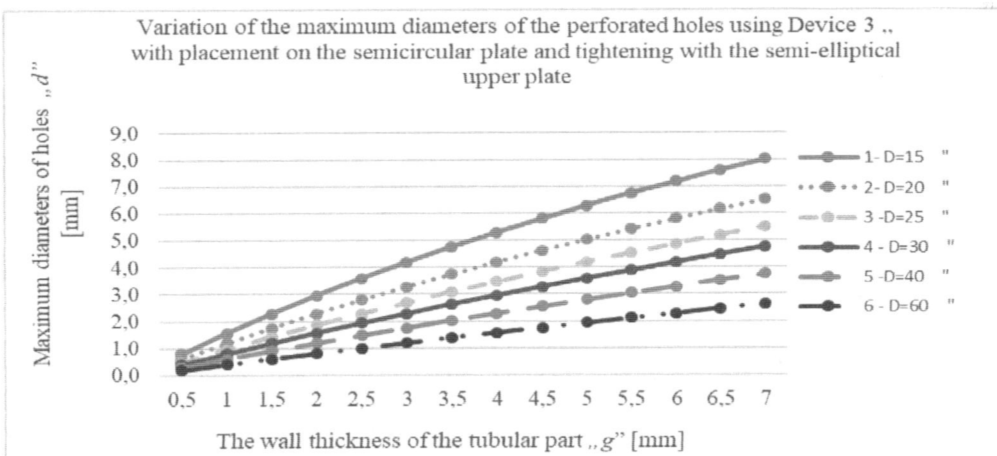


**Figure 9.** Diagram of diameters variation, by using the device 2, with placement on a semi-cylindrical plate and tightening with a semi-elliptical upper plate.

## 5. Comparative analysis of the values obtained for the diameters of the perforated holes for the three cases, respectively by clamping the part with the researched devices

A comparative summary table was drawn up for the results obtained by using the three proposed devices, for two types of pipes, with an outer diameter of 15 mm and 20 mm respectively, and comparative diagrams were drawn up regarding the values of the perforated hole diameters

## 6. Conclusions

1) Calculation relationships were obtained for the maximum diameter of the perforated holes so that the part does not deform, relation (6) for the case of device 1, relation (18) for the case of device 2, relation (12) for the case of device 3, depending on:

- the outer diameter of the pipe, D, respectively the radius R;
- pipe wall thickness, g;
- the characteristics of the pipe material, the breaking strength $\sigma_r$ and the permissible bending strength $\sigma_{ai}$;

2) Based on the calculation relations, the variation diagrams of the maximum possible diameters of the perforated holes were obtained, diagrams presented in figures 7, 8 and 9, which show the variation of the maximum diameters of the holes, depending on the wall thickness of the pipe, for a certain pipe diameter and certain material characteristics;

3) Based on the determined calculation relations, (6) (12) and (18) variation diagrams can be constructed for different other dimensions of round pipes and other mechanical characteristics of the material;

4) It is observed that the size of the perforated holes depends to a large extent on the thickness of the pipe wall, at the same material characteristics; the increase in the diameters of the perforated holes is proportional to the increase in the thickness of the pipe wall.

5) In figures 10 and 11 for comparison between the results obtained with the three types of devices, the curves of the diameters variation were presented, for pipes with an outer diameter D = 15 mm, in figure 10 and D = 20 mm, respectively, in figure 11; it is observed that the device 3, with the placement of the pipe on a plate semi-cylindrical and tightened with a semi-elliptical upper plate, allows the realization of the largest perforated holes, which means that it is the **optimal clamping device for use in single edge perforation of long pipes in air.**

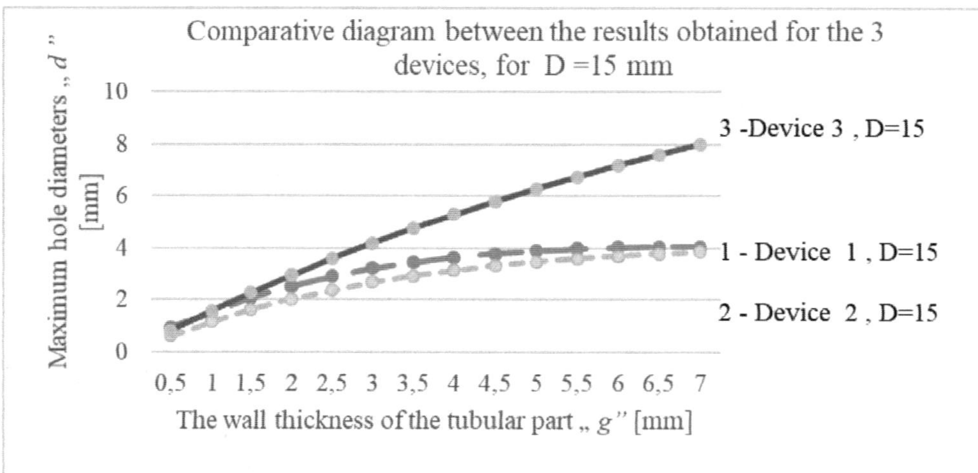


**Figure 10.** Comparison diagram of the variation curves of the holes diameters, with clamping the piece with the three clamping devices, for pipes with an outer diameter D = 15 mm.

Note: The curve 3 corresponds to the variation of the hole diameters in the case of the device noted 3
The curve 1 corresponds to the variation of the hole diameters in the case of the device noted 1
The curve 2 corresponds to the variation of the hole diameters in the case of the device noted 2

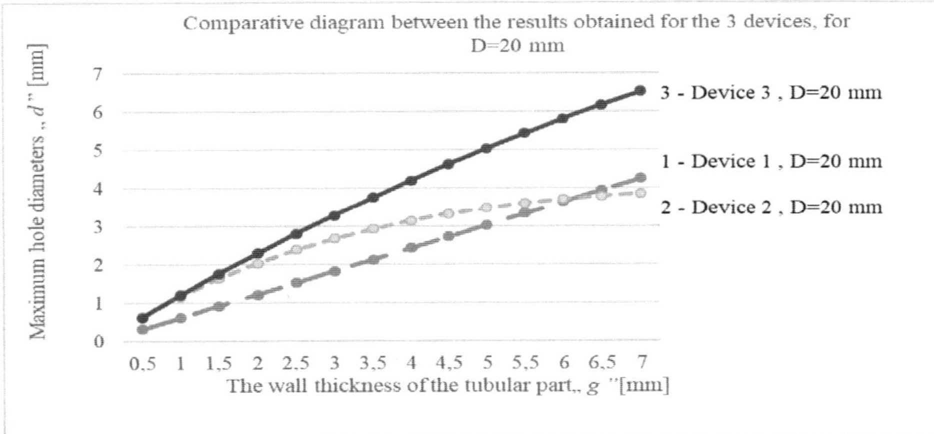


**Figure 11.** Comparison diagram of the variation curves of the holes diameters, with clamping the piece with the three clamping devices, for pipes with a diameter D = 20 mm.
Note: The curve 3 corresponds to the variation of the hole diameters in the case of the device noted 3
The curve 1 corresponds to the variation of the hole diameters in the case of the device noted 1
The curve 2 corresponds to the variation of the hole diameters in the case of the device noted 2

**Acknowledgment**
The paper is accomplished in the frame of the PhD research activity of the fist author, at the Politehnica University of Timisoara.

**References**

[1] Koç M 2008 *Hydroforming for advanced manufacturing*, Woodhead Publishing and Maney Publishing on behalf of The Institute of Materials, Minerals & Mining, (CRC Press Boca Raton Boston New York Washington, DC).
[2] Poiana D I and Ionel I 2020 Study of modern perforation processes specific to the automotive industry, *IOP Publishing-IOP Conference Series: Materials Science and Engineering* **89** https://iopscience.iop.org/article/10.1088/1757-899X/898/1/012043
[3] Kumar R and Hynes R J 2019 Thermal drilling processing on sheet metals: A review, *International Journal of Lightweight Materials and Manufacture*, https://www.sciencedirect.com/journal/international-journal-of-lightweight-materials-and-manufacture (accessed 01.06.2021)
[4] Borș A D 2018 *Doctoral thesis-Research and contributions on cold plastic deformation of magnesium alloy metal sheets* (original in Romanian: Teza de doctorat-Cercetări și contribuții privind deformarea plastică la rece a tablelor metalice din aliaj de magneziu) (Universitatea „Vasile Alecsandri" din Bacău)
[5] Gherghescu J et al 2007 *The interdisciplinarity-Theory and practice (*original in Romanian:Interdisciplinaritatea-Teorie și practică) (Editura Sfera)
[6] Drăghici G 2005-1 *The concept of mechanical processing processes (*original in Romanian: Concepţia proceselor de prelucrare mecanică) (Editura Politehnica, ISBN 973-625-224-8, Timişoara)
[7] Drăghici G 2005 *Integrated engineering (*original in Romanian: *Inginerie integrata) (* Editura Politehnica, ISBN 973-625-224-8, Timişoara)
[8] Marin C 2006 *Material resistance and elements of elasticity theory (*original in Romanian: Rezistenţa materialelor şi elemente de teoria elasticităţii), (Editura Bibliotheca, Târgovişte)
[9] Tripa P 1999 *Material resistance (*original in Romanian: *Rezistenta materialelor)* (vol. 1, Editura Mirton, Timişoara).

# AUTHOR INDEX